CORNISH MINING

CORNISH MINING

THE TECHNIQUES OF METAL MINING IN THE WEST OF ENGLAND, PAST AND PRESENT

Bryan Earl

D. BRADFORD BARTON LTD
TRURO

First published 1968 by
D. BRADFORD BARTON LTD
FRANCES STREET TRURO CORNWALL

Printed in Great Britain by
H. E. WARNE LTD ST. AUSTELL

CONTENTS

ILLUSTRATIONS

with nine figures, map and eleven other line illustrations in the text

PREFACE

MINING in the West of England carries a unique romance, and has such an ancient history that the techniques used make a fascinating study. A full account of all the known mining methods would require a book running into several volumes, and in a review such as has been attempted here the chief problem has been judging what emphasis should be placed on the various aspects of mining so as to give a balanced picture of the relative importance of their contribution to the main theme. Certain points have therefore only received brief mention, or have been passed over. The problems arising from divided mineral rights, for example, or the readiness of the Cornish engineers to adapt equipment to their needs, have only been noted, although playing an important part in explaining some of the "whys" of mining details. Also, some of the less usual machinery such as the engines that originated on the Continent using water rather than steam acting on a piston, to operate pitwork, unfortunately need a more detailed coverage to find a place. The layout and style of the larger illustrations is based on typical Cornish mine drawings, and along with the text are designed to try to capture some of the character of Cornish mining in the past.

My thanks for aid are due to many people: miners underground or tinstreamers at their frames, who courteously stopped for a "bit chat"—and many others. In particular I would like to thank J. H. Trounson, R. J. Law and H. L. Douch for information and helpful advice, the Miss Blights of Roskear for their kindness in helping me to obtain obscure information, and my publisher D. B. Barton for his encouragement.

BRYAN EARL

Robin Hood's Barn,
Merryhill Green,
Wokingham, Berks.
June 1967

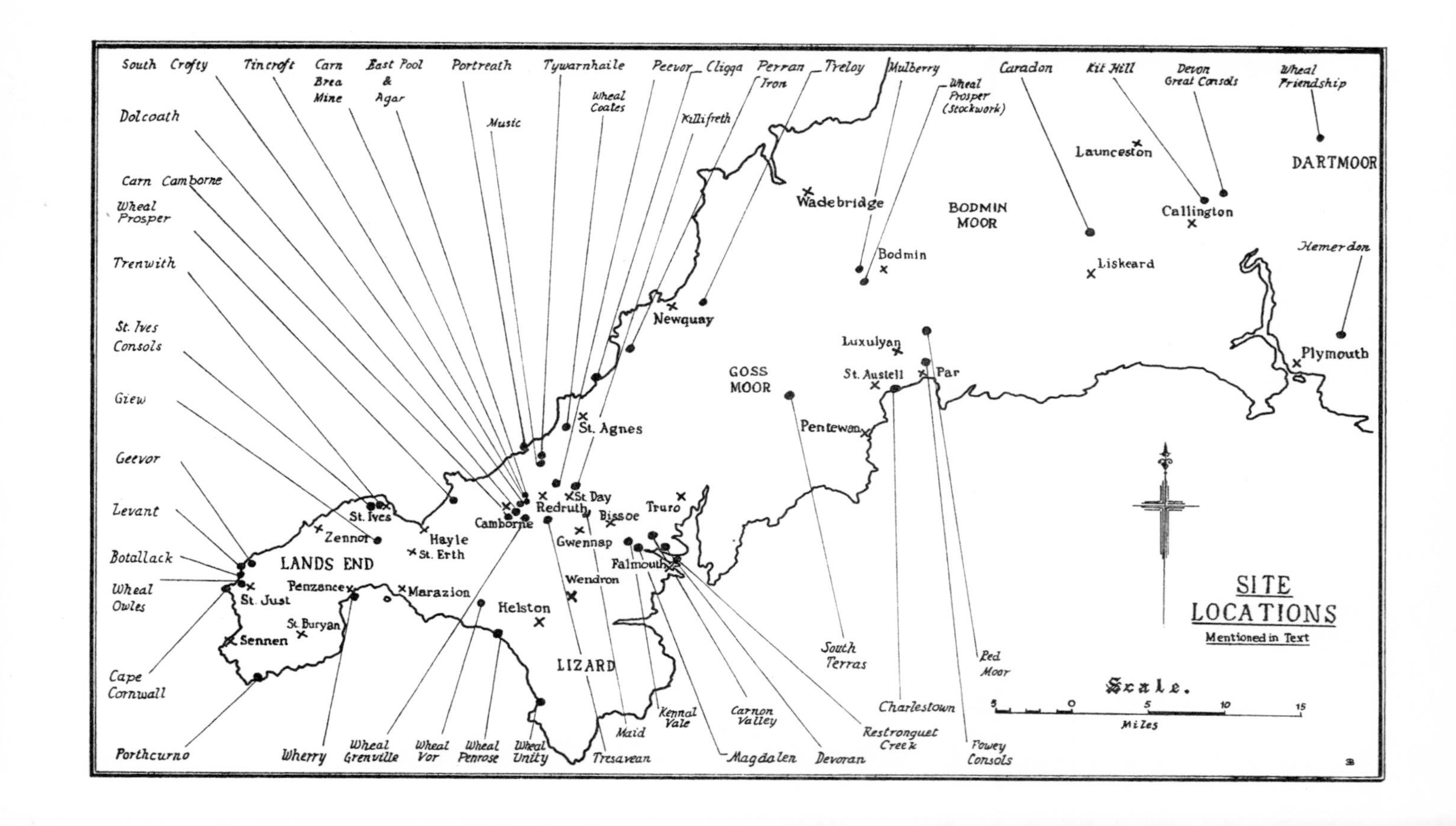
SITE LOCATIONS
Mentioned in Text
Scale.
5
0
5
10
15
Miles
South Crofty
Tincroft
Carn Brea Mine
East Pool & Agar
Portreath
Music
Tywarnhaile
Wheal Coates
Peevor
Cligga
Killifreth
Perran Iron
Treloy
Mulberry
Wheal Prosper (Stockwork)
Caradon
Kit Hill
Devon Great Consols
Wheal Friendship
DARTMOOR
Launceston
Callington
BODMIN MOOR
Wadebridge
Bodmin
Liskeard
Hemerdon
Plymouth
Newquay
Luxulyan
St. Austell
Par
GOSS MOOR
Pentewan
St. Agnes
St Day
Redruth
Bissoe
Truro
Camborne
Gwennap
Falmouth
Wendron
Helston
LIZARD
St. Ives
Zennor
Hayle
St. Erth
LANDS END
Penzance
Marazion
St. Just
St. Buryan
Sennen
Dolcoath
Carn Camborne
Wheal Prosper
Trenwith
St. Ives Consols
Giew
Geevor
Levant
Botallack
Wheal Owles
Cape Cornwall
Porthcurno
Wherry
Wheal Grenville
Wheal Vor
Wheal Penrose
Wheal Unity
Tresavean
Maid
Kennal Vale
Carnon Valley
Magdalen
Devoran
Restronguet Creek
Charlestown
South Terras
Red Moor
Fowey Consols

INTRODUCTION

HERODOTUS, in the fifth century B.C., refers to the ancient Phoenicians trading in tin which originated in the "Cassiterides"—a group of mysterious islands situated vaguely in a sea beyond Europe. Remains have been found in Cornwall dating local tin smelting operations to about the same period as Herodotus, thus supporting the assumption that Cornwall could indeed be the famed and mysterious mist-wreathed 'tin-islands'.

Cornwall has a most ancient and romantic history of mining. It is impossible to trace just how, when and where the metals originally were discovered, but mining has played an important part in the West of England as far back as records can be followed. Although the main fields were in Cornwall, Devon held some most important deposits, both of tin and copper, particularly in the Tamar district. Tin is usually thought of as being the principal metal mined in the West of England, but from the beginning of the eighteenth to the end of the nineteenth centuries copper was also of great importance. A wide range of other lesser minerals was also mined to a smaller extent at times, such as lead, silver, tungsten, arsenic, zinc, nickel, cobalt, bismuth, antimony, uranium, iron, manganese, pyrite (for sulphur), and fluorspar.

At first the ores were won on a small scale but as the demand for metals grew, with consequent good prices, the outputs increased accordingly and by the middle of the nineteenth century the West of England mining region was supplying nearly half of the total world output of both tin and copper, and this only fell with the rise in production from cheaper foreign sources. Records of outputs prior to the nineteenth century are incomplete, but they do help to give a useful estimate of activity. The relative prices paid for tin and copper have varied greatly over the years—tin sometimes being less valuable than copper, whereas at present, it is approximately twice the value of copper (in 1870 black tin was £70 per ton and copper £90 per ton).

Laws relating to the searching for, and mining of, the ores were introduced in Cornwall as in other ancient mining regions and 'Stannary

Courts' (comprised of tinners) were held to administer these laws. From time to time, the full 'parliament' of these tinners would meet at the desolate Crockern Tor on Dartmoor.

The land which held the valuable minerals was owned by 'lords' (either owning the rights for themselves or for the Duchy of Cornwall) or by the Duke of Cornwall, and 'dues' or 'dish' had to be paid by the miners to the lords or to the Duchy. These dues were calculated as a

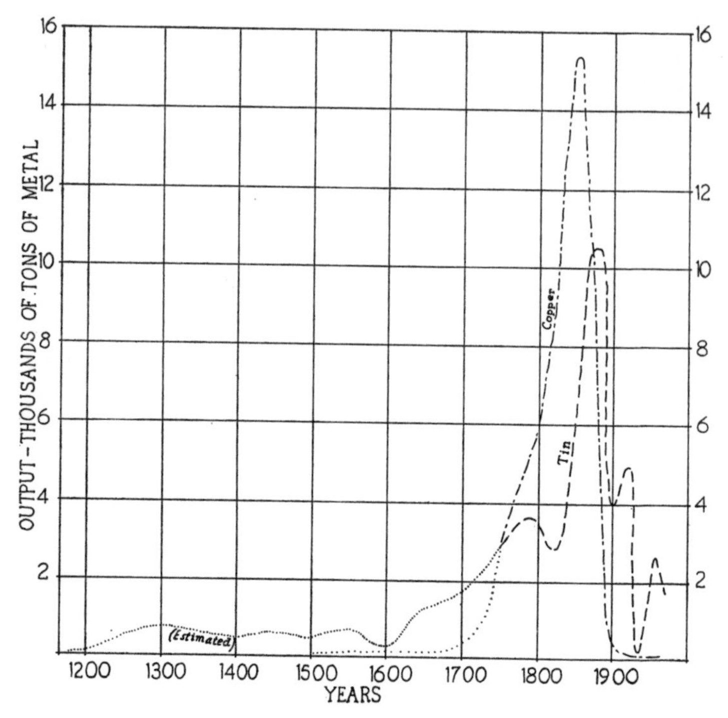

GRAPH SHOWING ORE OUTPUT OF THE WEST OF ENGLAND, 1200 - 1900 A.D.

fraction of the value of the mineral won and varied according to circumstances. Thus a lord might charge dues of $^{1}/_{15}$th of the value of tin ore extracted during the start of a mine, abating this to $^{1}/_{18}$th if a steam pump was installed or other heavy capital outlay incurred. This was done in order to encourage the mining 'adventurers' (those who put up the money to finance the venture) to win ore on a large scale and so benefit the lord. The dues for copper were frequently higher— $^{1}/_{8}$th to $^{1}/_{5}$th. The land was divided up into areas, or 'setts', and mined to the boundaries of these according to the agreement reached between the lords and the adventurers.

The early workers were usually family groups, who found the capital (often in the form of their own skills) necessary to build the required plant. The concentrates of ore they produced were either sold to a smelter or reduced by the miners themselves, the tinners having to wait for a 'coinage' time before they could be paid. Later, as the mining operations grew in size, 'cost book' companies were formed—each a company of unlimited liability into which the shareholders either paid 'calls' (or demands for money to finance the operations) or shared profits, according to the prosperity of the mine. The mine also kept a 'cost book' which recorded in concise detail the expenses, dues, and earnings of the miners and much other information. Usually, a total of 64 or sometimes 128 shares (sometimes later sub-divided) were held by the adventurers who had a wide variety of interest, the majority being speculators, mineral lords and merchants. At least once a year, but usually every 12 or 16 weeks, the books were examined at the mine and at these times dinners were held and a large amount of punch, ale and brandy was consumed. Beautiful dinner services which were used at the mines on these occasions are still in existence. Towards the end of the nineteenth century, limited liability companies were being set up and the 'cost book' system gradually came to an end.

Grand dinners also accompanied the principal method evolved to sell the ores (particularly copper) to the smelters, known as 'ticketing'. At fixed dates, the smelters' agents sampled the ore piles which accumulated at the mine, and offers were then made in writing on 'tickets', at an agreed date, the highest bid taking the ore, and a fine lunch was provided for the smelters' agents. For a time, this practice spread to the selling of tin ore but has now given way to less clumsy business methods.

The overall authority for the running of a large mine was held by the 'Captain', 'Agent', or more recently Manager, who was appointed by the adventurers or directors. He had 'underground captains'—knowledgeable and experienced miners—who were responsible to him for supervising the underground work, and 'grass captains' overseeing surface operations. The 'purser' (often one of the adventurers) kept the accounts, receiving and paying out monies. The workers were drawn from far and wide, many having to walk several miles to and from the mine daily. The usual time spent underground—'spell' or 'core'—prior to the 1820's was six hours, but this was later extended to eight hours. Some mines were tolerably pleasant to work in, being cool and well ventilated, but in others where the temperature rose to over 100°F, with very high humidity and poor ventilation, work was arduous. In the 'ends' the air was sometimes so poor that the candles would only burn if held nearly horizontally, and were snuffed out whenever possible, as "a candle took as much air as a man". The miners had set 'croust' times when they gathered at a certain spot to rest, have a 'morsel' of food (latterly this was frequently a hot pasty wrapped in a towel and sent down in an old wooden 'diny', or dynamite, box) and a chat over a pipe.

Apart from those engaged at the mines, many were employed in ancillary work as consulting engineers, mechanical engineers, workers in the foundries, rope makers, candle makers and a host of other occupations. Because of the widespread nature of the mines and the relatively small scale of the work when compared with the intense concentration of industry in the Midlands, Cornwall never suffered the over-crowding and squalor that characterised the factories and dwellings of the Industrial Revolution elsewhere.

The region has a very large number of mines (approaching 2,000) varying in size from small exploratory adits to an enormous ramification of tunnels—60 miles or more—extending down to 3,000 ft. beneath the surface, in at least one case. At the time of the greatest activity, between 1820 and 1870, about 400 mines were active, with outputs varying from one ton or less of concentrate per month, to 80 tons or more for a large mine such as Dolcoath. The extent of mining is indicated by the estimates made of those employed by the mines in the year 1837-1838: 18,472 men, approximately 5,700 women and 5,700 children. The average monthly wage was 52s. 6d. for men and 14s. 6d. for women and children. During the same time, approximately 56,860 tons of coal, valued at £48,381; 14,056 loads (of 50 ft.^{3}) of timber, valued at £36,545; 300 tons of gunpowder at £44 per ton, and 1,344,000 lbs. of candles at £35,000, were used. Cornish mining has had many cycles of booms and depressions each stimulating the development of mining techniques which have benefited not only mining but also engineering in the widest sense.

CHAPTER ONE

LODES AND ORES

THE MODES of occurrence of the ores in the ground of the West of England, and the nature of the minerals making up these ores, have been important factors in the development of mining methods. The geology and mineralogy of the area is full of interesting material for study.

Before the upheavals which resulted in the formation of the lodes, the area comprising what is now the Isles of Scilly, Cornwall and Devon, was covered with sedimentary rocks, compacted by their own weight. The sedimentary material had come from older rocks, which had been broken down by weathering. About two hundred and fifty million years ago (after the Carboniferous Period) some of the changes in the earth's structure, due to contraction, caused masses of molten rock to be forced up under great pressure from below into the sedimentary rocks. This 'magma' of granite rock intruded more in some places than others. Near this hot, new rock, charged with reactive gases and solutions, the sedimentaries were baked and altered, the amount of alteration depending on the composition of the rock, its chemical reactivity and the distance from the hot magma. Today minerals caused by this baking (such as white needles of andalusite) can be found in some of the sedimentary rocks, showing that they were in this region of change—the "metamorphic aureole"—produced by the granite magma. Andalusite slate can be seen in a patch in the cliffs, walking along Porthleven beach towards Wheal Penrose, as one nears the granite. To the Cornish miner, the sedimentary rocks, in and out of the metamorphic aureole, are known as 'killas'.

The more silica (SiO_2) there is in a rock, the more 'acid' it is. Granite is an acid rock, made up of three minerals—quartz, which is free silica, orthoclase felspar, and a mica. The quartz appears as shining, transparent and glass-like crystals; the felspar forms opaque white crystals, and the mica bunches of thin shining leaflets. According to the impurities which were present in the original magma, other 'accessory' minerals may be present. Tourmaline is a typical accessory, usually occurring as black needles among the other minerals making up the granite. The rocks on the beach at Porthcurno show good crystals of tourmaline, and the quartz, felspar and mica can be readily seen. Tin (and frequently copper) ores are usually found associated with acid rocks, such as granite. As the content of silica falls, the minerals making up a rock change, and the rock progressively becomes more 'basic'. A good pocket magnifier, about ×10, is useful to have when out prospecting, enabling the forms of the minerals to be identified.

As the granite magma cooled, it contracted, cracked and moved, along with the surrounding killas. Some of the fissures produced were filled with quartz porphyry material to form hard rock barriers—elvan dykes—which can be difficult to work through when mining. Later, in certain areas emanations of hot gases and solutions from the magma were forced into other fissures and weakness planes. These mineral-bearing solutions came from certain restricted localities in the still liquid core of the granite, known as 'emanative centres'. The solutions were charged with the elements which were deposited forming the minerals which make up the valuable ores. As the mineral fluids moved up and out, they met changes in temperature and pressure. Particular minerals were deposited to make a lode in a fissure, where the temperature and pressure conditions were favourable. This gave a sequence of mineral zones, as the distance increased from the emanative centre. Because of this, tin ores are often found deeper in a lode than copper ores, and lead and silver ores in turn are usually further from the centre than copper. Thus the 'high temperature' tin-copper area of St. Agnes has a fringe of 'low temperature' lead-silver, mined at a greater distance from the emanative centre which appears to have been located under a line joining St. Agnes to Cligga Head.

In many districts in the West of England, the main bulk of the granite itself holds small amounts of the valuable minerals, as accessories. Thus crystals of cassiterite, the tin ore, may sometimes be found in the granite—the mud on some of the side roads between St. Austell and Bodmin ground up from the local rock can assay a pound weight of tin per ton.

There were several periods during which mineralization occurred. The fissures, which were becoming lodes where they contained deposits of valuable minerals, were altered, and complicated in the minerals they held. Further changes due to subsequent earth movements caused more fissuring and weakness planes to develop in the rocks. Minerals continued to be deposited, and the lodes already formed were distorted and faulted. Due to there being several periods of mineralization, the favourable localities for the minerals varied, causing telescoping, overlapping and complication of the deposits. As a result, in some areas tin and copper ores are predominant, while in other areas it is lead, and in some there is a mixture. Also, there were periods when solutions were introduced which reacted with the materials already in the lode, changing the minerals from one to another. Boron solutions have frequently changed original lode material to tourmaline. The later mineralizing solutions often made 'crosscourses' (or 'guides')—complex fissures filled with clay ('fluccan') or minerals of little value, and cut across valuable lodes, which can be the cause of water trouble, as they afford a passage for water from the surface, old mine workings, or the sea.

By about two hundred million years ago (during the Triassic age) the mineralizing processes had stopped. A feature of the way in which the valuable lodes were formed is the tendency for particular ores to be deposited in lodes which run more or less parallel to one another. This "preferred" direction of lodes varies slightly from district to district, so that in the Camborne area tin and copper ores are usually found in lodes running in an east-west direction (actually E 20°N) whereas the local silver/lead lodes run north-north-west. In the St. Just area, the tin/copper lodes usually run north-west. The miner working one lode makes use of this, searching the rocks on either side, by tunnelling or drilling out at right angles to try to find more ore. A 'caunter' lode is one whose direction does not conform with others in the locality.

Many granite areas in the world have little or no valuable ore deposits connected with them, as no ore-forming elements have been associated with the rock magma. The West of England is one of the areas fortunate in having wide-spread mineralization. The mineralization processes conform to those found in most other metallogenetic provinces throughout the world. For its area, Cornwall has a rather large variety of minerals and forms of ore deposit.

By the end of the period of lode formation, the chief areas of mineralization were those around St. Just, Camborne, St. Day (Gwennap), St. Agnes, St. Austell, Caradon, Callington and Dartmoor. The richest was in the Camborne-Redruth district, and was connected with the Carn Brea and Carnmenellis granite masses. The surface of the ground has since been greatly altered by erosion. Continuing convulsions in the earth caused some of the land to sink beneath the sea, and then be raised out again. Some of the killas covering the solidified granite intrusions was eroded and removed, in places exposing the top of the granite which was in turn broken down. The lodes held in the killas and granite suffered denudation at the same time. Between Camborne and Hayle, at various view points on the road, the remains can be made out of the flat marine platform to the north of Carn Brea, cut away when the area was under the sea. The broken down rocks and the lodes contained in them formed new layers of pebbles and sand, or were deposited in stream beds and valleys. Before the granite intrusions and mineralization, Cornwall and Devon, as they are now, appear to have been part of a large continent. The granite intrusions formed a core, which became the backbone of the present land remaining after most of the area had sunk beneath the sea.

The Lizard is an unusual district—for Cornwall—as the rocks are of a 'basic' type, containing less silica than the main granite intrusions and more magnesium based minerals. Tin is not found to any extent in the Lizard; but traces of metals which are usually associated with basic rocks, may be detected. Copper, which sometimes occurs in basic as well as acid rock, has been mined in the Lizard area, for instance at Wheal Unity, near Mullion Cove. The copper was often found 'native' in the metallic state, and not as a compound.

Due to the great variety of elements which were injected into the rocks, many of the minerals making up the 'country rock' around the lodes were changed by chemical reactions, or altered by the effects of pressure or shear stress. Iron was nearly always a common element in the solutions—and resulted in the rocks near the mineralized areas often looking 'dirty' and stained brown. This can be a useful lead when prospecting, and the ochre formed from the iron in many mines gave a prevailing rusty appearance to anyone who had been into the mine.

The present coastline of West Cornwall is revealing, as the sea is breaking down rocks which hold many lodes. The exposure of lodes in cliffs is not a common feature in the world, and the cliffs near St. Just are particularly interesting. Here the sea has excavated into the weaker rocks near the lodes, to form inlets—zawns—and brilliant splashes of colour, such as the green caused by the copper ores reacting with the saltwater spray, are an impressive sight.

The discovery of the useful materials held within the earth required inquisitiveness and observation. Metals which can occur in the metallic state, such as native copper, were no doubt utilised first. The art of extracting metals from ores which are non-metallic in appearance needed a considerable development of knowledge and skills. Tin is a metal of this class—the commonest ore is the heavy oxide: cassiterite. To obtain metallic tin, the ore must be freed from waste minerals and reduced to the metal by heating with materials such as charcoal. Similarly, although copper does occur native, the main sources are of more or less complex compounds, such as the sulphide chalcopyrite, a mineral made up of copper, iron and sulphur. A relatively complicated smelting process is required to reduce such ore to copper. Although it is probable that the ancient inhabitants of Cornwall knew of tin and copper prior to 500 B.C., the most likely early systematic approach to mining appears to have come from the people of the Mediterrranean area, who had a developed civilisation and knew of metallic ores as well as how to win and smelt them. Explorers from the Mediterranean came to Cornwall, and, discovering the valuable ores, started mining and trading in them.

The early workers found their ore in the surface deposits which had resulted from the breakdown and weathering of the original lodes. Some was dug out from the portion of lodes which were exposed at the surface—or "worked into the backs"—where water solutions had permeated into the minerals, decomposing them and making the ground soft and easy to remove. Such weathered deposits at the top of a lode are known as gozzans, and can be distinguished from the surrounding surface, being coloured brown-red by iron minerals, and of a cindery, porous nature. Such upper parts of a lode were sometimes called a 'broile' or 'bryle'.

Due to the lack of good, readily made metal tools, most of the early work had to be in lode stuff which had been broken away by weathering, and washed into beds, either in existing streams or where there had been an old watercourse. The action of running water tends to concentrate most metallic ores—the heavy values of, for example, tin or lead being deposited as the stream of water washes away the lighter waste rock particles. Other localities were on plateau land—where rock debris, derived from granite containing sufficient disseminated cassiterite as an 'accessory', had been concentrated into a workable deposit.

A metal miner has to be both mineralogist and geologist, to be able to recognise the ores, and to deduce from the rocks what the likely pattern of mineralization may be. In Cornwall a knowledge of mineralogy is particularly useful, as the lodes and ores can be most complicated, and sometimes hold minerals which are unexpected, rare and valuable. The 'old men' developed great skill in being able to interpret the characteristics of the deposit they were working from the appearance of the ground, and deduce what would be the best action to exploit the lode. A piece of lode material crushed up and swirled with water on a shovel told the miner, by the positions taken up by the mineral particles, what values were held by the rocks.

Unfortunately, some misconceptions developed over the years, and are still heard amongst the miners. Typical sayings are that "no good tin is found in the granite" and that "the best ore lies in lodes in the region of the granite-killas contact". Although these theories may be right for a given area due to the zoning effect on the deposits, depending on the distance from the emanative centre, they do not hold good for the entire West of England mining region. The ideal location for an ore mineral in relation to the rock type varied from centre to centre, and was chiefly controlled by the temperature and pressure gradients existing in the area at the time of deposition. In the past such misconceptions affected prospecting, resulting in faulty deductions.

Some special terms have evolved which are used to describe the positions of the lodes. As shown in fig. I the *strike* is the main direction of a lode along a horizontal surface, usually given as a compass bearing. The *dip* is the angle of depression of the lode, measured from the horizontal surface, and at right angles to the strike. The *underlie* is the angle the lode makes from the vertical, at right angles to the strike. The *hanging-wall* is the ground above the lode, the *foot-wall* the ground beneath the lode. The valuable material is nearly always unequally distributed in a lode: some parts of a lode may be rich, others poor. A *shoot* refers to the vertical extent, a *course* the horizontal extent, of values in a lode.

The actual lode structure varies from place to place and the widths differ considerably. In the St. Just mines the lodes were rather narrow, typically about 1 ft. 6 in. to 3 ft., whereas in Camborne they were up to 20 ft. or more wide, but the overall average was about 3 ft. 6 in. In places great masses of ore may be found in a block of ground, such as the great 'carbonas' found in St. Ives Consols, just to the west of St. Ives town. A typical carbona was a mass of rich tin-impregnated granite about 100 ft. long by 30 ft. high and wide.

In places the ore may be found in a 'stockwork'—a zone of 'country' with numerous thin, mineralized veins, or sprinkled throughout with spots of ore, possibly 50 to 60 ft. wide, the rock usually being somewhat altered and softened due to decomposition by the mineralizing solutions.

Fig. II is a section across a representative Cornish lode, one worked by South Crofty at Pool near Camborne. The solutions and gases

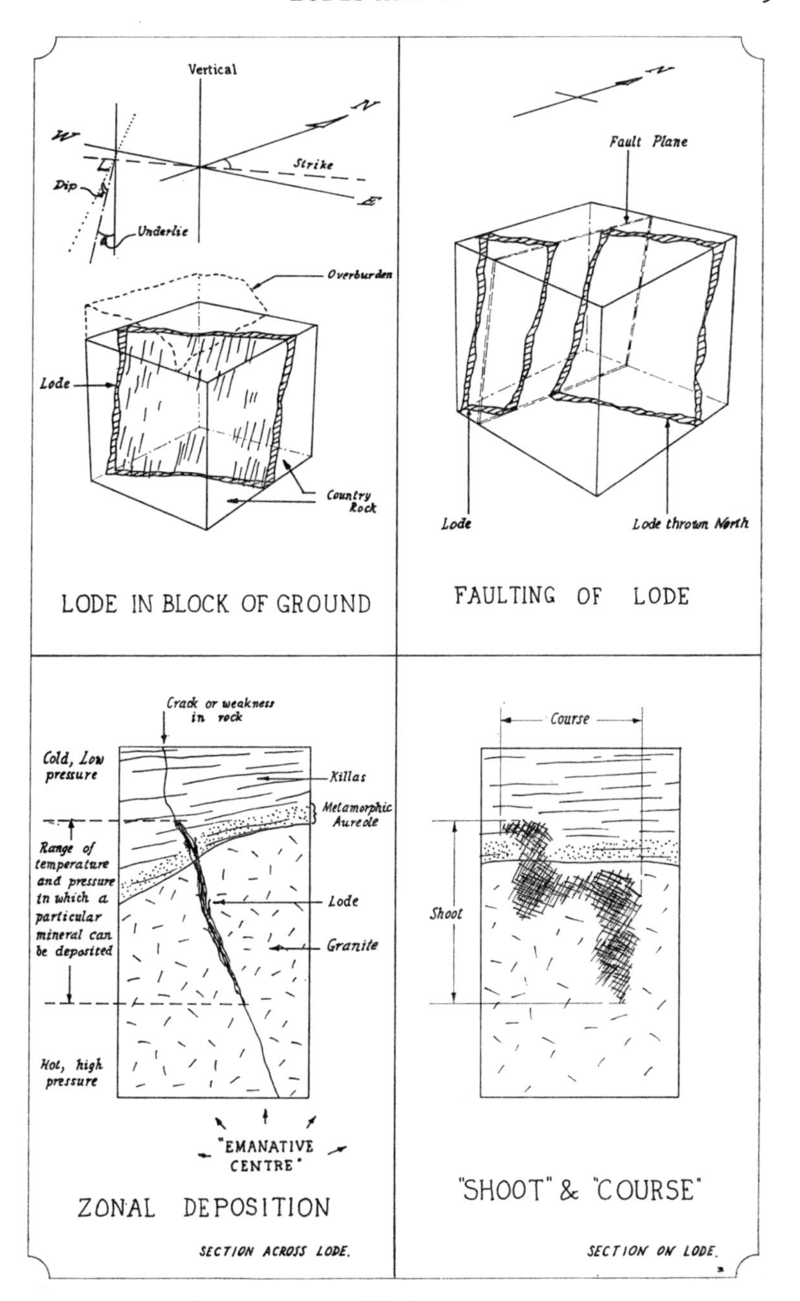
Vertical
N
W
Strike
E
Dip
Underlie
Overburden
Lode
Country Rock
LODE IN BLOCK OF GROUND
N
Fault Plane
Lode
Lode thrown North
FAULTING OF LODE
Crack or weakness in rock
Cold, Low pressure
Killas
Metamorphic Aureole
Range of temperature and pressure in which a particular mineral can be deposited
Lode
Granite
Hot, high pressure
"EMANATIVE CENTRE"
ZONAL DEPOSITION
SECTION ACROSS LODE.
Course
Shoot
"SHOOT" & "COURSE"
SECTION ON LODE.

FIG. 1

which have passed through the lode channel over the ages have produced changes seen in both the granite country rock and the actual materials held in the lode, such as forming the 'peach' and tourmalinising the granite. 'Peach' is the miners' description of the green lode-filling material often found in Cornwall, made up of a felted mass of

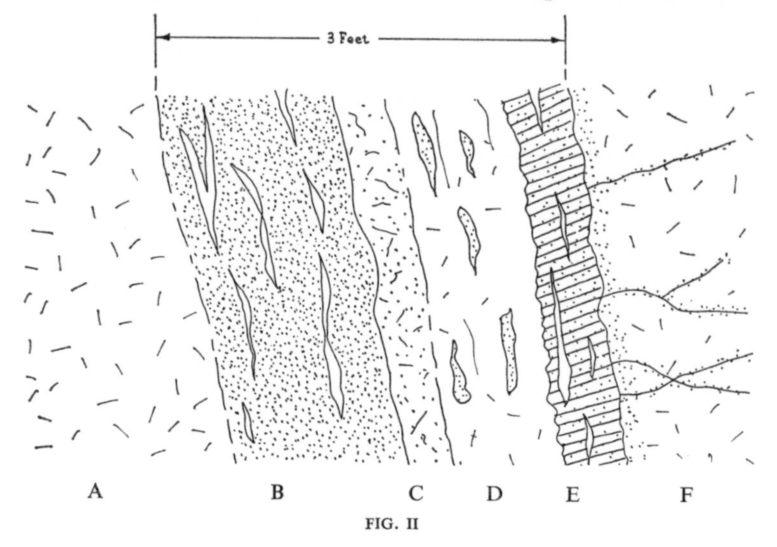

FIG. II

A Granite, tourmaline impregnated near lode walls.
B Hard dark green peach with quartz veins.
C Quartz, peach and arsenopyrite.
D Quartz with streaks of chlorite peach.
E Soft peach with quartz veins.
F Granite, tourmaline impregnated near lode walls.

crystals of the minerals tourmaline and chlorite. This lode is worked for tin, the cassiterite being deposited over the lode width in varying concentration. The cassiterite crystals are frequently very small and difficult to see and are often 'locked up' in the crystals of the waste minerals or 'gangue', such as quartz, but tend to be larger nearer the surface than at depth, sometimes forming 'bunches' or rich parts of nearly pure ore—a fine sight for the miner. A small bunch is known as a 'squat'. The mineralized wall-rock is known as 'capel'. The more locked up the cassiterite crystals are in the gangue, the more crushing is needed to free the ore and enable it to be concentrated for smelting. Fig. III is of a typical crystal of cassiterite (tin oxide, SnO_2). The shapes of crystals found in the lodes vary considerably. There may be much modification of form, and the mineralogist learns to be skilled in identifying the different habits. Pure cassiterite is white, but is nearly always found in shades of brown, or black. The ore may be given several names, such as 'tin' (although it is not tin metal), 'black tin' (concentrated cassiterite), or 'cassiterite', the mineral name. Wood-tin and toad's eye tin are particular forms the ore may take when deposited in certain fashions. Occasionally empty spaces called vughs or druses are found in lodes. They are often filled with carbon dioxide gas, and frequently lined with fine crystals of the minerals present in the lode.

The content of tin in a sample is usually given as pounds weight per ton. At the moment (1966) with the present costs of running a mine,

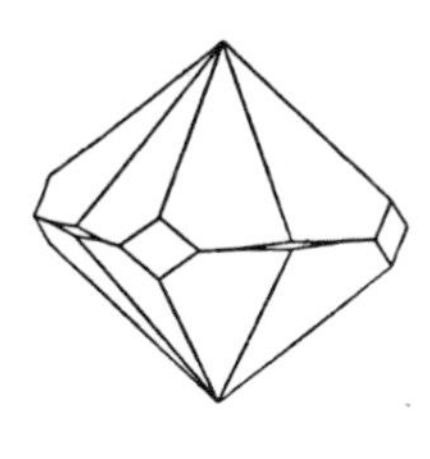

FIG. III
CASSITERITE

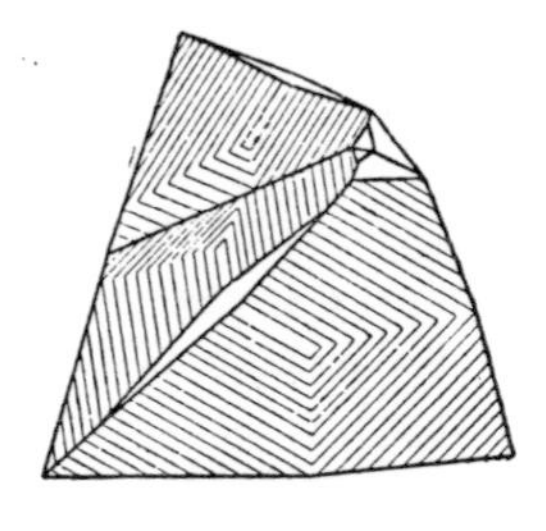

FIG. IV
CHALCOPYRITE

the methods used, and lodes of reasonable width, it would be profitable to work ore holding slightly less than 20 lb. of 'black tin', or tin oxide, per ton. The price received for the concentrate has not only to cover the actual mining of the ore, concentration and profit, but also the whole working of the mine, including wages, pumping, development work, machinery, depreciation and taxation. With the present price of tin (about £1,300 per ton), lodes which in the past were too poor to work—or 'below the limit of payability'—are now beginning to be good propositions, although costs of working have of course also increased.

Copper was won in great quantities in the West of England, 'yellow copper ore' being the main source. In order to be profitable, the ore had to be proportionately richer than tin, as copper was cheaper. The yellow copper ore was normally quite easily seen, and at times, such as in Devon Great Consols, the lode was virtually pure ore, with little waste. The ore itself was chalcopyrite (a copper-iron-sulphide $CuFeS_2$), rather brittle, heavy brass-yellow crystals found in various forms. An important feature in the dressing of copper ore was the hand picking to remove waste, which was readily seen. Another important ore of copper was chalcocite or redruthite, so named because it was found in quantity in that area. This grey, 'vitreous copper' ore was particularly rich, its chemical composition being Cu_2S. It occurred in a compact massive form, and was formed in the lodes by a process known as secondary enrichment. Where the lode came to the surface, water charged with various materials in solution percolated down, causing changes in the composition of the lode. Lode material was dissolved by this water, and reacted with the other minerals in the lode and in solution. At suitable points the new compounds which had been formed crystallised out—the exact location of the point depending on the characteristics of the component. In this way the original copper ore—chalcopyrite—was dissolved, carried down, and re-deposited as the new mineral redruthite. Because the conditions under which the re-crystallisation could occur were rather limited, the concentration of the new mineral in that zone tends to be high. Many elements carried in the lodes have been concentrated by this secondary enrichment process. An interesting example was the mysterious "Green Jim" uranium lode of South Terras mine in the Fal valley about 5 miles west of St. Austell. The original pitchblende, or primary ore, a mixture of black uranium oxides, was changed to the bright green secondary ore mixture of torbernite (complex phosphate of copper and uranium) and autunite (complex phosphate of calcium and uranium). This formed magnificent green and yellow leaflets in a lode which was worked in the 1900's for its radium content, one of the elements resulting from the radioactive breakdown of uranium over the ages. The operations were surrounded by a cloak and dagger atmosphere, which encouraged strange rumours as to the effects of the radioactivity.

An important by-product of many operations was arsenic, which was refined to white arsenious oxide. This was released when many of the ores had to be roasted to break down the mispickel or arsenopyrite (Fe.As.S 'silver mundic', or lölingite with less sulphur) and other materials which interfered with the concentration process, such as pyrite—FeS_2—or 'mundic'. If the arsenopyrite was left unchanged, it was concentrated with the tin, and gave a poor product which incurred penalties from the smelter. A few mines were opened to work arsenopyrite alone when there was a great demand for arsenic for dusting the cotton crops in the United States to combat boll-weevil. Lead and silver have been of some importance—usually occurring together in the heavy blue-black cubic ore galena, a mixed lead/silver sulphide of varying composition

PbS/Ag_2S. The ore is quite massive, and was separated by gravity methods from the gangue. Iron has also been mined in Cornwall, usually as the red oxide, hematite Fe_2O_3, the yellow-brown oxide, limonite, approximately $2Fe_2O_3.3H_2O$, or the carbonate, siderite $FeCO_3$. The Perran iron lode near Perranporth was an example of open-work mining.

Cornwall is notable for the wide range of minerals, some of which form valuable ores, which occur along with the main ores of tin and copper, or in isolated lodes. Many of these have been worked from time to time when reasonable concentrations have been found. Examples are wolfram (ore—tungstate of iron and manganese [Fe.Mn]WO_4) used to make special alloy steels; zinc (ore—sphalerite or 'black jack' ZnS, zinc sulphide) used in alloys and galvanising; nickel (ore—Kupfernickel NiAs) for alloys; manganese (ore—pyrolusite, MnO_2) used at the time in glass-making; antimony (ore—stibnite, Sb_2S_3) used to harden lead; cobalt (ore—smaltite, cobalt arsenide $CoAs_2$) to make blue glaze for chinaware, and alloys; bismuth (ore—bismuthinite, bismuth sulphide Bi_2S_3) for low melting point alloys; and gold, found in small nuggets in tin streaming operations, particularly in the Carnon valley.

CHAPTER TWO

SURFACE WORKING

THE EARLIEST operations for winning ore in the West of England were on the surface tin deposits. How long ago these tin workings started has been difficult to assess—the ground having been worked over so many times that material and objects which could date the sites have been inter-mixed, thus confusing estimates of a historical sequence. Tin metal remains have been found in a smelting pit in the St. Just area which can be dated back at least to 300 B.C., and it appears that tin working commenced in the West of Cornwall considerably before that time. Even as late as the seventeenth century, three quarters of the ore produced in Cornwall came from surface deposits.

Several different methods of winning the tin developed over the ages. It is most probable that the first ore was found in the stream beds where the ground had been broken up and washed along by the running water. The movement of water was an effective way of concentrating finely comminuted ore stuff, as the heavy tin ore, being more difficult to move than the lighter waste (sp.g. cassiterite 6.8-7.1; quartz 2.65), tended to settle into any pockets in the stream bed, or at any irregularity, such as a bend or boulder, where eddy currents could swirl up the rock particles and promote the conditions for separation. Searching along the river beds for these concentrates was the earliest form of 'tin streaming'. The method used was to find a favourable site and dig out the deposit, which was then shaken and stirred carefully in a stream of clear water. The heavy ore was settled as much as possible, and as this was done, the gangue material gently raked off. The cassiterite remaining was then scooped out. It is probable that the old workers did this in large wooden bowls, as is done to this day by Malayan tin streamers. An improvement on this technique was to dig out the deposit into a stream of water, diverted along a channel provided especially for the dressing process, slowly raking up against the current of water. The ore was carried furthest against the stream because of its high density, and over a period a concentrate was formed and then removed. The concentration of the original deposit of stream tin was such that the ore was relatively easy to work up to a grade fit for smelting. Normally the values were free of minerals that would cause trouble in smelting, due to water having leached out the more objectionable impurities, such as pyrite and arsenopyrite, and this resulted in a preference for stream tin by the smelters when lode ores started to be worked. Lode ores were sometimes contaminated and had to be specially treated to remove the impurities.

The demand for tin encouraged exploration, and development of the less easily found surface deposits. Due to the breakdown of the rocks after the period of mineralization, a large amount of ore was laid down in surface detrital gravels in areas which became moorland, and was also carried into the beds of streams which later dried up. The valuable material was usually covered by further rock debris, holding little tin. Very few of these moor deposits are now left in their original condition, the 'old men' having worked them over time and time again. The main areas of these sites, moving east from Land's End, were: Penrose, near Sennen, and in Bosworlas streams about 1 mile S.E. of St. Just; streams around Tregadgwith, Bojewans and Kerris near St. Buryan; in the Penzance area along the valleys running inland from Newlyn, particularly inland from Trereife (where a very old rag and chain pump, operated by a waterwheel 12 ft. diameter by 8 in. breast, was found about 200 years ago), and on the Drift moor; in Marazion marsh, where there was much trouble in pitting due to the remains of the submerged ancient forest there. Old moor and stream tinners have also been active inland between Morvah and Zennor, and along valleys in the Towednack area, near St. Ives.

Alluvials in the St. Erth river valley were exploited, especially between Relubbus and Carbis, and the deposits near St. Erth church were recently treated by hydraulic sluicing and gravel pumps. Near Helston moor and stream deposits have been worked between Trenear and Porkellis. The Carnon valley deposits had extensive workings inland from the extremity of the navigable branch of the Fal. Originally these were true alluvial workings, but later streaming collected values from the St. Day mines' tailings. Deposits under Restronguet Creek have also been extracted. The stream tin works at Treloy, near St. Columb Minor, was notable for containing rings and brooches of Roman origin. The Goss moors have been extensively turned over by tinners.

The St. Austell district had a large number of stream tin works, several of which were active, using old dressing methods, into this century. The chief localities were along the valley inland from Pentewan, and in the Par area stream tin was worked to the shore. There were works down Luxulyan valley and on Red Moor, Levrean and the surrounding moors. Deposits have been treated on Bodmin Moor, the Kit Hill district, and Dartmoor, where the very ancient operations are believed to have produced a large amount of tin.

The tinners were so active in many places that the unworked ground was known as 'whole-ground'. The richest alluvial deposits were those of central and western Cornwall. The tin bearing ground was normally found from five to thirty feet below the surface, and often in two layers split by a band of clay, the lower layer being the richest. A typical cross section of a moor deposit (at Red Moor, in the St. Austell district) was made up of 2-3 feet of peat and 1-3 feet of granite gravel resting on 4-5 feet of tin-bearing gravel, consisting mainly of granite fragments with some flints. A thin scattering of cassiterite, often rounded fractured crystals, was found in the lower rich layer, which rested on a bed of brownish clay. Pockets in the bed were often rich in tin and known as 'whirls'.

The values in the detrital deposits varied considerably, and the lowest concentration which could be treated depended on the depth of the overburden, the situation, and the skill of the tinner. For example, the Happy Union stream works at Pentewan in the early years of the nineteenth century worked gravels holding 11¼ lb. per ton which was considered a good proposition.

The moor and old river bed material was usually coarser in size than that found in a running stream, where the exposed concentrate had been subjected to more break-up by water movement. Large pieces of cassiterite-bearing rock, which had been broken away from a lode but had not suffered prolonged attrition, were known as 'shoad stones'. Later, miners used these stones as a guide when prospecting for lodes, as they could be followed back to the parent lode by noting how they were placed compared with the slope and contour of the ground.

At the time of Carew (*Survey of Cornwall*, 1602), the land was divided by law into two categories: Several and Wastrell, or enclosed grounds and common. In the laws relating to tin working, both surface and lode, these terms were defined as follows:

> "In *Severall* no man can search for tin without leave first obtained from the lorde of the soile, who, when any myne is found, may worke it wholly himself, or associate partners, or set it out at a farme certain, or leave it unwrought, at his pleasure.
>
> In *Wastrell* it is lawful for any man to make trial of his fortune that

way, provided that he acknowledge the lorde's right by sharing out unto him a certaine part which they call toll: a custome savouring more of indifferencie than the tynner's constitution in Devon, which inable them to digge for tynne in any man's ground, inclosed or unclosed, without licence, tribute or satisfaction. The Wastrell works are reckoned amongst *chattels* and may pass by worde or will."

From the fourteenth century, the prospective tinner had to mark the extent of his claim by 'bounding' with some indication—such as digging a hole in the ground—which had to be renewed each year or the privilege was lost. Copper claims were not subjected to this 'bounding'. Today, the 'mineral rights'—the minerals held in the ground divorced from surface rights—may be held by the person owning the land, or by the Duchy of Cornwall; thus to obtain the right to work a piece of ground (or 'sett') requires the knowledge of to whom an application must be made, and it is often difficult to trace the necessary details due to ancient complex arrangements.

In the past, the streamer or moor tinner applied for the working rights of the ground he was proposing to examine. Next, small rectangular pits were sunk and, if in unstable ground, these were lined with timber so as to prevent the walls from collapsing. The tinner examined the various layers of the deposit exposed in the pits. If good 'stuff' was found, further pits were sunk to test the extent of the deposit and so determine the location of the water channels that would be required. A leat was then constructed in the lowest lying part of the workings (known as a 'level'—not to be confused with a level mined in rock), and the material was eventually washed into this leat and concentrated. Sometimes another, higher leat was made to direct a flow of water over the ground and so wash the gravels down into the lower stream or level. Working up from the level, the tinners removed as much of the barren overburden as possible, and allowed the good 'stuff' to be sluiced down, wheel-barrowing any difficult ground. If the gravels were heavily water-bearing, the tinners would not employ any sluice water, but would bail, or use a rag-and-chain pump driven by water-wheel or windmill so as to expose the ore. Pitting alone might have to be used if the tin stuff was in flat ground, working out as much of the tin-stuff as possible, and the hillocks left by the workers as they threw them up from cast to cast were known as 'shambles'. In this way, the tin-ground was systematically turned over.

The waste 'tailings' that were discharged from the levels caused the rivers and estuaries to be silted up for several miles in some areas, and this resulted in the tinners being held in disrepute. The Hayle estuary to this day is still partly silted as a result of these old workings. The cassiterite particles found in these detrital beds were often relatively coarse (up to approximately ¾ in. diameter), smooth water-worn pebbles, which had originally been large crystals or portions of 'wood-tin' deposits. The material washed into the lower level was hand picked to take out these large pebbles and shoad stones, which were then crushed to free the tin, the powdered material being added to the bulk to be washed clean. Originally, the crushing was done by hand. The ore was put in hollows in granite blocks (or mortars) and beaten with hand-held stones. If the cassiterite was in fine crystals, and well locked up in the gangue, the material was crushed even finer in a 'crazing mill' made up of two stones, one turning on top of the other on a vertical axis, in a similar manner to a flour mill.

Due to the confused state of these old workings, it is difficult to trace when more elaborate treatment methods were developed. It is probable that the moor and stream deposits were worked over several times before sorting and crushing were applied, and enterprising individuals found that there was a living to be made from old turned-over ground, provided this was suitably dressed. The old tinners were often family groups, who worked in the isolation of the moors. The reputation of these 'old men' for ingenuity and for an independent spirit has lived on, and Cornish miners even today are notable for the same traits of character.

An interesting example of the complicated methods required to work one of the more extensive detrital deposits occurred at Restronguet creek, where the rich tin-stuff lay in a bed covered by clay and water. At first, the water was dammed off from part of the bed and this was then stripped down to the tin gravels. Later, shafts were sunk into the deposit. The most extensive working of this nature in the 1800's used a shaft made up of iron cylinders, bolted together, the first section having a sharp edge. Sinking through the river bed was accomplished by mooring heavily laden barges on to the top section at high tide, the weight of the barges settling down with the ebbing tide ramming the cylinder down through the mud. A second shaft, equipped with a Cornish pump, was sunk in the bank at the side of the creek and the two shafts then connected by a level, further levels branching out into the tin gravel. All the workings had to be well timbered for support and as the tin ground was removed the supports were allowed to collapse. The impervious clay layer over the deposit prevented too much water for the pump from entering the levels. The two shafts, in conjunction with doors in the levels, allowed air to circulate. Earlier attempts at mining in the creek bed had only used one shaft, and the poor ventilation from the lack of air circulation permitted inflammable gas, from decomposing organic remains in the clay, to accumulate to such a point that it was occasionally ignited by the miners' candles, causing several scorchings. Inflammable gas (methane) was not usually found in Cornish mines, as the rocks do not hold organic material that could decompose to methane.

The ability to work into the actual lodes was controlled by the availability of tools, but where the lodes had been weathered to gozzan, extraction was relatively easy. Simple wooden shovels, or, later on, wood metal clad, were adequate to dig away the cindery, cellular material. Gozzans could be readily seen and aroused the interest of early prospectors. The ground was stained brown and had few or no plants growing on it due to the high concentration of toxic minerals. The pits made by working down into a gozzan were called goffins, coffans or coffins, and such workings are often very old. The loose material was dug out and either shovelled up step-by-step to the surface (shammelling), or hoisted out of the trench by a bucket and windlass, or if in the side of a hill or cliff, wheel-barrowed out to be treated. Although the gozzan might extend to a considerable depth, coffin excavations were not normally very deep—up to 50 ft.—due to the difficulty of supporting the sides. The remains of these open trenches, some of considerable size, can still be seen near St. Agnes, for instance at Wheal Coates, and also on the site of West Wheal Jane, about ½ mile north of Bissoe. Again, a deposit may have been worked at several different times. As tools and methods were improved, the tougher rocks outside the gozzan zone could be attacked and hard-rock tunnelling and underground mining started.

A special open mining system was developed to work the wide stockworks found in the West of England. These were mostly for tin but a few, such as Wheal Music (two miles north of Scorrier), were for copper, or other metals such as the Hemerdon wolfram-tin stockwork, near Plympton. The cassiterite in a tin stockwork was found in a wide band of rock—averaging 60 ft.—with a rather low metal content, a deposit of 5-10 lbs. tin per ton being considered workable. The rock, usually granite, in the stockwork band was sometimes altered and broken down to a soft nature by the mineralizing and other solutions which had permeated the zone. The felspar of the granite was frequently partly changed to kaolin (china clay) and forms of mica and other alteration minerals, making the ground easy to break. At first pits were opened in the ground and the ore hoisted out by bucket, but later as the 'quarry' face became larger new methods were developed. Levels were prepared at the bottom of the pit running ahead of the working face. The levels were given a timber roof and the material broken off the face was allowed to fall on to these timbers. Small wheelbarrows or waggons on rails were brought under the broken ore, and this was run through the timbers into the waggons and taken out of the pit. In many cases the level extended to a valley which had a stream driving the dressing plant. The workings left spectacular excavations in many cases. A famous stockwork mine, which is traditionally said to be very old, is the open pit at Mulberry, 1½ miles N.W. of Lanivet, near Bodmin. The mineralization is rather unusual, as the strike of the Mulberry deposit runs 'caunter' or at right angles to the predominant Cornish east-west tin strike. Another large open pit can be seen near Lanivet—Wheal Prosper—adjoining the old Bodmin road. In later working, some mining was done from horizontal levels driven into the face to take out rich patches, and for a short time in the 1930's and 1940's lorries were used in some of the sites, doing away with the old timbered level system.

The working over of old dumps left from previous mining activities has been a source of ore since earliest times. The 'old men' often left material in waste dumps which it was uneconomic or impossible to treat, since they lacked the later skills and knowledge. With the changing values of metals, and improved treatment, what was formerly waste could become a paying proposition. Old dumps could also be a source of ores which, at one time, were of no value—such as the pitchblende in the dumps of Wheal Trenwith, near St. Ives—and which later became valuable. The old burrows were sampled and picked over and, if results justified it, the dressers obtained the rights to treat the dump, and removed it to their dressing floors. In the old days this was effected by wheelbarrow or horse and cart, but today modern power shovels, excavators and dump trucks are busy on this job. In the 1890's there were many small dressers, particularly in the Camborne-Redruth area, treating old burrows. They also frequently dressed ore for small workers, who could not afford the machinery, payment often being taken as a percentage of the values extracted from the ore. Today, the equivalent dressing floors have machinery to cope with a wide range of material from several different sources, combining the ores at suitable points in the treatment. Thus the large stones from an old dump are stamped to a 'pulp' in water, and mixed with the fine stuff coming in from river sand, and so on. Some surface detrital deposits are being re-treated again today and dressed at central plants such as at Bissoe.

With the spread of mining activities, and the start of deep mining, large amounts of 'waste' were discharged from the floors of the mines into the rivers and streams. A new form of tin streaming started, which is sometimes confused with the winning of tin from the true original detrital deposits, and which had to be more skilful as the mine orestuff contained more impurities. The new streamers took this mine waste, and re-treated it, to extract as much as possible of the residual values which had escaped the mines' plant. In dressing lode ore, it was usually necessary to crush to a fine powder in water (pulp) and work on this to extract the cassiterite. The finer the cassiterite crystals were in the gangue the finer the ore had to be powdered to set them free, and the finer the pulp the more likely the fine tin would be carried over in the waste. The overall economics of mining controlled just how much treatment the ore could be subjected to, as it is impossible to recover every last pound. Thus the waste 'tails' could be worked on by skilled men to give up further concentrates of value, the economics of the operation being changed as the raw material was often there virtually for the taking. Tin streaming of this nature occurred in every mining district of any size. The streamers were most skilled workers, devising ingenious equipment often from the simplest materials such as furze, old sacking and "home made" plant built to treat the fine 'slime' tin. Ways of splitting the waste streams to achieve the maximum recovery of pulp went hand in hand with their machinery. As mineral dressing methods improved, and the mines brought in more efficient techniques, so the tin-streamers developed new ways to extract what was left to them. Today tinstreaming of this nature can be seen in several parts of Cornwall, for example along the Red River valley between Camborne and Redruth. The present works treat waste dumps and true river deposits as well as mine refuse. A few minutes spent talking to the friendly twentieth century tin-streamer soon lets one into some of the mystique of Cornish mining, and quickly supports the legendary stories of the individuality and character of the 'old men'.

The beaches of many parts of Cornwall hold tin, either in the silt from the outlets of rivers which drain former mining districts, as at Gwithian, or from lodes which outcrop in the sea bed. The tin values occur as a rather dilute layer in the sand, and are worked by removing the ordinary sand and carting the tinstuff for treatment. A small plant, extracting about one ton of black tin a month, was formerly situated on the beach near St. Agnes. Similarly, beaches were worked near Cape Cornwall, where after storms particularly rich sand was found which had been brought in by the heavy seas. At present, beach sand is being treated, especially near St. Agnes, and dredging operations are in progress off Hayle on such deposits.

Copper has been recovered by a process known as cementation. Water which held copper salts in solution was made to pass through a pit which was filled with old iron. The copper was brought out of solution and replaced the iron which was dissolved. Some of these cementation pits held a wonderful collection of old iron pots, kettles, bedsteads and shovels, all gradually being changed to copper. The ironmongery was dug over from time to time to expose as much to the solution as possible, and eventually taken out (a clean knife blade put into a copper solution soon shows a coating of copper, the iron being displaced in accordance with the positions held by these two elements in the electrochemical series). Usually about a third of the iron was dissolved and replaced by copper. The water pumped from the Devon Great Consols copper mine was treated in this manner, as was the water from the Great County Adit draining the mines in the Gwennap district, the cementation pits being near Bissoe. The water discharged from the pits at Bissoe was led into large ponds where the air oxidised the dissolved iron, which was precipitated as ochre and found a ready market.

Surface methods of winning ores in the West of England have always been on a relatively small scale. An average present day Malayan tin dredge will treat approximately 16,000 tons per day of detrital tin bearing gravel, holding about 1 lb. per ton, whereas an equivalent Cornish streamer would treat possibly 5-20 tons per day. However, estimates of the tin won in Cornwall from surface workings indicate that over one million tons of tin metal were extracted by the moor and stream tinners over their long history.

MINERS IN AN UNDERHAND STOPE. No. 3 branch lode at 12 level, in Geevor. The jackhammer drill is putting down shotholes to blast the ore, which the miners are standing on, with the footwall to the left and the hangingwall to the right. The lode is about 2 ft. 6 in. wide at this spot, and the gunnis left where the ore has been removed can be seen behind the miners. [1956]

DRILLING IN A STOPE DRIVE IN GEEVOR, looking down from a main level (No. 12) at miners. Note the jackhammer machine attached to an airleg, with air and water hoses, and the chain ladder used for access down the shallow winze. [1956]

DRIVING A DEVELOPMENT LEVEL 'ON LODE'. Miners drilling shotholes with a drifter machine, mounted on a hydraulically positioned boom. The machine is about to start drilling one of the 'cut' holes, the stemmer partly inserted in a hole to the left being used to aid alignment. Note the lode crossing the face at about 15° underlie. Taken in No. 9 level at Geevor in 1956.

CHAPTER THREE

WINNING GROUND

THE INCREASE in the value of tin, together with the development of metal tools, resulted in the exploitation of the lodes in the hard rocks. Prior to the sixteenth century, work which required breaking hard ground, or going to any depth, had only been undertaken to a limited extent. The "winning of ground" or ability to expose the ore in the rocks, had always been a major problem in Cornish mining. 'Hard rock' mining owes much to Cornwall, where tools and methods, some imported from abroad, were developed or improved. By the nineteenth century, the West of England was one of the world's major mining fields, both in skills and size.

There were two main problems standing in the way of large scale lode mining: the flow of water into the workings as soon as any depth was reached, and the hardness of the rocks. As these were overcome, other difficulties arose for the engineers. The early machinery for deep mining was massive, and the remoteness of some of the mines, and the position of others perched on the edge of cliffs, called for great skill to install such machines. As the depths of the mines increased, the methods of hoisting the ore and aiding the descent and ascent of miners had to be improved. Geological knowledge was required to understand the complex and faulted lodes, with shoots and courses of values. Problems from heat and poor ventilation became more apparent as the mines went deeper, the temperature of the rocks increasing with depth (in Cornwall about 1°C for every 200 ft., depending on the district).

The early miners levered and wedged away the rocks with simple tools. 'Plug-and-feathering' was one of the first methods used: a conical rod of metal was driven between one or two thin metal wedges put in a rock crack, or later in holes drilled with a rectangular bit, so forcing the rock to split. Hard bronze and crude iron were the tool metals. Shovels were of wood, later tipped with metal to improve wear. Any metal was expensive and soft by today's standards. Iron was forged from the spongy metal and slag produced by the old reducing furnaces. So precious were these early tools that the miners tied a cord from them to their waist so that they would not be lost in the dim oil, tallow or candle light. The moils and gads of present-day Cornish mining are descendants of these early tools. Many special forms of pick and hammer were made to work the different widths of lode.

Fire-setting was a very ancient aid to breaking rock. A wood fire was built against the rock, and allowed to burn for a considerable period, so heating and expanding the rock. The fire was then allowed to go out so that the rock cooled and shrank. Sometimes water was dashed on it to speed the cooling, but in a tunnel it was often virtually impossible to reach the fire due to smoke and heat. The expansion and subsequent contraction caused the rock to crack and flake off. This was a slow process and filled the workings with smoke, but some remarkable mining was done and fire-setting was still in use up to the beginning of this century in some European mines.

Quicklime was also used to break rock. The lime was put into a hole bored about 3 in. in diameter, some water poured in, and rammed tightly with clay stemming. As the quicklime was slaked to the hydroxide, great heat was developed, expanding lime and water and so splitting the rock. Remains have been found of lime breaking, for instance, at Wheal Magdalen, near Ponsanooth.

Adits were used in Cornwall well before 1500 to drain off water which ran into the workings. A small tunnel was driven, usually into the outcropping ore, at the lowest possible point such as the bottom of a valley or cliff. This tunnel "to the daylight" was an adit. A slight slope upwards from the daylight end was aimed at so that the water would drain from the lode into the tunnel and so away. Lodes make channels in the rocks which water can seep through. and Cornish mines, with their steeply dipping lodes, always have had to face a serious water problem. Where the lodes were in flat ground, pumping was used to keep the water down to expose as much of the lode as possible. A variety of methods were employed: bailing with a bucket and windlass was the simplest. The wheel pump was an arrangement resembling an overshot waterwheel but working in reverse, picking up the water and tipping it out at the highest point. Another early device was an endless chain of pots, arranged to scoop into the water and tip out at the surface. Rag and chain pumps were much used and were made up of an endless rope or chain with bolts of rags, leather or such material, tied on at intervals. One side of the endless loop ran through a pipe made of bored-out logs, or logs cut in half, hollowed and bound together to make a pipe. This pipe ran down into the water and at the top discharged into a channel away from the work. The chain of tight fitting rags was pulled up through the pipe by running over a wheel. This was worked through gearing by men, horses or waterwheel, the mine water being drawn up the pipe in the segments between the rags.

The bucket pump was more elaborate and effective, made up of a plunger, known as a 'bucket', worked up and down in a pipe by rods. The bucket was made to fit the pipe closely—usually by leather packing—and had holes covered by a flap which acted as a valve. A similar flap valve was built into the pipe, near the bottom, and when the bucket was pulled up, it sucked the water up past the flap of the bottom valve, the bucket's valve remaining closed. On the downstroke of the bucket, the bottom valve closed, the bucket's valve opening as the water was forced through it and above the bucket. On the next upstroke more water was drawn into the pipe through the lower valve, to be pushed above the bucket on the downstroke. In this way the water was lifted up the pipe in a series of 'plugs'. These pumps were frequently worked by a crank on the end of a waterwheel axle. Several pumps could be placed at different levels down a shaft, one pumping up to the next. The driving waterwheel might be below ground, the pumped and spent waterwheel water running to an adit. By discharging into an adit, power could be saved in proportion to the depth from surface. Sometimes the workings were bailed out with a bucket, or a 'water-barrel', which could be automatically emptied at the surface, on the hoist rope. This method is used to this day if an extra flood of water enters a mine.

All of these aids to the winning of ground are very old, and fire-setting, rag-and-chain and bucket pumps were perhaps known to the Phoenicians and Romans. By the sixteenth century their use was well established in the Cornish mines, but during the next hundred years, the mining problems stimulated imaginative minds, and an extraordinary time of invention commenced.

Until the sixteenth century, mining appears to have attracted but little interest from those not directly involved. Accounts of costs of working, and laws relating to mining can be found, but "everyone knew" who needed to, how to mine. It was not the fashion for literate and educated men to describe such affairs. However, the importance of mining was soon to be recognised, and handbooks and descriptions of mining began to appear. Agricola's *De Re Metallica*, published in Basle in 1556, gives extensive and detailed accounts of mining methods, and Carew, in his *Survey of Cornwall*, published in 1602, describes Cornish mining methods of the 1500's which agree very closely with Agricola's.

In 1698, Savery, a military engineer, invented a water pumping system which depended on the pressure and vacuum that could be obtained from steam let into closed vessels. His patent shows that he saw in the drainage of mines one of the main uses of this invention. Savery's pump may have been tried at Wheal Vor, or one of the other Godolphin mines, but if so, it was not successful. Thomas Newcomen, an ironmonger and tool merchant of Dartmouth, about whom little is known, learned of the water problem on his visits as a merchant to the Cornish mines. After many years of trial, he designed the first steam engine—a remarkable feat, and one of the most important advances in engineering. As Savery held the master patent on 'Raising of Water . . . by the Impellent Force of Fire', Newcomen had to go into partnership with him. The first known Newcomen engine was at a colliery near Dudley Castle in Staffordshire, and built in 1712. However, by 1715 a Newcomen engine had also been erected at Wheal Vor by the Godolphins—who were frequently associated with trials of new methods which could improve the working of their mines.

This type of engine was made up of a vertical cylinder and piston connected to a beam, which was pivoted at its centre like a see-saw, the other end of the beam being coupled to bucket pumps in the shaft by chains and wood rods. Steam at low pressure (about one or two lbs. per square inch above atmosphere) was let into the cylinder, and the piston allowed to rise. Cold water was then sprayed into the steam, condensing it and establishing a partial vacuum. The pressure of the atmosphere pushed the piston back into the cylinder, thus pulling down the beam (hence the alternative name of 'atmospheric' engine). The other end of the beam pulled up the rods to the bucket pumps, raising the water. This work cycle was then repeated. Counterweighting of the pump equipment in the shaft—the pitwork—was soon added: one end of a pivoted beam, similar to that of the engine, was coupled to the rods, the other end of this beam having a counterweight box filled with stones. This balanced the parts of the pump and engine, and made the whole work more smoothly and efficiently. The early Newcomen engines were coupled to as many as six pumps in the shaft (each pump lifting up to ten fathoms), and each of them had a separate connection to the engine beam. Later, the pumps were coupled off one main rod moving up and down in the shaft.

Wheal Vor was sunk 60 fathoms below adit with its Newcomen engines. By 1778 over sixty were recorded in the West of England. In 1765, James Watt repaired a model Newcomen engine, and this led him to develop his steam engine. Watt greatly increased the efficiency by adding a steam condenser outside the cylinder, and using steam rather than atmospheric pressure to do work on the piston as well as provide a vacuum. The condensing of the steam inside the Newcomen cylinder caused a great loss of energy, as the cooled cylinder took up heat from the steam on each cycle. Watt teamed up with the great Midlands' entrepreneur Boulton, and began installing Watt engines in Cornwall by 1778. A brilliant engineer, William Murdoch, was employed by Boulton and Watt and was in charge of the erection of their engines in Cornwall. The Watt engine was important to the Cornish mines, as it used considerably less coal than the Newcomen, and coal was expensive as it had to be brought from Wales: the Watt engine requiring about one third of the coal to pump a given quantity of water that a Newcomen engine did. Boulton and Watt charged the mines a 'premium' on their engines, which was calculated on the savings from the new engine over the atmospheric for pumping a set amount of water. This led to litigation, there being two lawsuits, one being pressed to an issue.

In 1811 Trevithick first applied high pressure steam (40 lbs. per square inch) to a Watt type engine at Wheal Prosper (Watt engines ran on steam at 2 to 4 pound per square inch). Also, the steam valve was adjusted to cut-off the steam early, allowing the steam to expand and do work over the remainder of the piston stroke. Watt had previously tried to use steam expansively in this manner, but had not achieved any economy as his pressures were too low to take advantage from this effect. The high pressure proved to be considerably more economical than the low, and the system gradually spread, evolving into the 'Cornish' engine. Great care was taken to lag Cornish engines with non-conducting materials to save heat, and the valve mechanism was designed to give very close control over the steam. The size of these engines was given as the bore of the cylinder, such as the "90 inch at New Cook's shaft" (South Crofty).

The actual pumps down the shaft remained the bucket type, working on the upstroke, until about 1810. After that time plunger pumps, working on the downstroke, were used except for the extreme bottom 'lift'. Bucket pumps continued to be used there due to the facility with which a bucket pump could be lowered when shaft sinking and the ability to repair the bucket, or put in a 'drop clack' if the lower valve failed—the valves would not be accessible in a plunger pump if water flooded over them. The water came up the rising main pipe of each lift in the shaft into a cistern or small reservoir at the next pump. On the upstroke, the plunger or 'pole', which was connected to the main pump rod running from the engine, drew up water from this cistern into the cylinder in which it worked (the polecase) via a suction flap valve. On the downstroke, the plunger pushed the water out of the polecase, pressing shut the valve to the cistern, and forcing open another flap valve leading to the next section of rising main up the shaft. Due to the layout, the valveboxes were known as the 'door pieces' (the water was drawn from the 'windbore' in the 'cistern' through the 'clack' at the 'H piece' to the 'polecase', then was pushed through the clack in the 'top doorpiece' standing on the H-piece and up the 'rising main'). The valves were known as clacks from the noise they made on seating. Water can only be 'sucked' up about 25 ft., controlled by the pressure of the atmosphere which actually forces the water up in a suction-type pump. Water can be mechanically 'pushed' as high as is necessary. Plunger-pole pumps were particularly effective in the deeper mines and could deal with minewater which was loaded with suspended rock particles. The steam engine raised the main rod, the actual pumping being done by the weight of the rods on the downstroke. In most cases, the engine was placed close to the shaft, but sometimes the power was taken (as with some waterwheels) by means of 'flat-rods' to a distant shaft. Using pivoted arms, the movement was transferred to horizontal rods running on rollers, between the engine and shaft. A second pivoted coupling, or angle-bob then worked the pump rods in the shaft.

The efficiency of these engines was quoted as a 'duty'—millions of foot pounds work done per bushel of coal burnt. The speed of the advance in engineering at this time is shown by the improvements in duty. In 1813 a good engine had a duty of about 20, whereas by 1834 it was nearly 100. By 1900 Cornish pumps were draining mines up to 500 fathoms deep (Dolcoath). The first half of the nineteenth century was the period when Cornish engineering flowered. Men such as Woolf, West, Grose and Sims designed better and finer machines, and engineering works started in Cornwall to meet the enormous expansion in demand from the mines. Trevithick assisted the development of steam power by his advocacy of 'high steam' and invention of the Cornish boiler. Foundries such as those of Harvey's at Hayle perfected techniques to permit the use of high pressure steam, and brought in machine tools to do the precise machining required on the new engines. In later developments, compound engines were designed using various arrangements of cylinders in attempts to expand the steam more efficiently in stages, although only a few on this principle worked in Cornwall. The great pride taken by the engineers was shown in the beautiful proportioning of the parts, and their ornamentation. This pride spread through the mines, from the miners' ability to bore well placed shotholes quickly, to the workmanship on the buildings.

By connecting the beam to a crank with a 'sweep' rod, beam engines were made into 'rotative' engines (an idea not worked out originally in Cornwall), and used to drive machinery and work as whim or hoisting engines, replacing the less powerful waterwheel and horse—the old 'horse whim' thus being replaced by the 'fire whim'.

A working Cornish pump was indeed a sight to see. The great stone building, with its arched doors and windows, stood right at the heart of the mine. Going inside, one was met by the neat, turned woodwork furnishings, and the clean sweet smell of the hot oil characteristic of a working steam engine. The enormous cylinder, usually clad with varnished wood bound with polished brass, stood in the middle. The brightwork of the valvegear, spotlessly polished, twinkled as the scoggan catches smoothly operated the valves. Led up the staircase by a proud engineman, one came to the floor, the 'middle chamber', with the great cylinder cover and nozzles, the shining piston rod disappearing up into the complex of the parallel motion and main beam. One more trip up the balustraded stairs to the bob—the huge beam, swinging quietly up, pausing, and then majestically sweeping down again. At the 'nose' of the beam was hung the first of the pump rods, probably of 18 in. square pitchpine, disappearing all the way down the shaft below—perhaps more than 2,000 ft. The bob paused at the top and bottom of the stroke to control the rate of pumping (timed by the cataracts), and to allow the valves to seat all the way down the shaft. The whole engine was remarkably quiet, and all that could be heard was the soft, deep "Boom" as the steam rushed into the condenser, quickly followed by the sound of fresh steam entering the cylinder through the steam valve, the clicking of the valvegear and the flow of water into the pan of the air pump, at each cycle. Coming down and out of the enginehouse, one might see a sudden brilliant orange glare, as a firedoor on the boilers was opened to let a shovel full of coal be thrown on the fire, just "so", in such a way that the coal landed in the right spot to give an even firebed. Of steam, smoke or noise there was practically none.

The last of these wonderful engines on a Cornish mine, at Robinson's Shaft of South Crofty, stopped work in 1955, after 101 years of active life. Now, electrically driven multi-stage turbine (or centrifugal) pumps placed in stations in the shaft, do its work. Electrical pumps were first used in Cornwall about 1906 during the period known as the "electric boom" in a shaft of Tywarnhaile Mine, near Porthtowan. Usually they have to be made of special stainless steel alloy to resist the corrosive mine waters (the large pipes and cast iron valves of the Cornish pumps were sometimes even lined with wood to protect them from corrosion, and hippopotamus hide used for facing the clacks, to resist destruction by acid mine waters). For the high lifts in mines, the turbine pump is made up of a series of discs, turning in a casing, both of which have suitably shaped waterways cast and machined in them. Water is fed in at the centre of a disc and 'spun' out, kinetic energy being given to it. The water coming out of the edge of one of these impeller discs is guided by the casing into a diffuser where the kinetic energy is converted into pressure before being guided back to the centre of the next impeller, a sufficient number of stages being used to achieve the pressure required to carry the water up the desired lift. In this way a great volume of water can be lifted out of the mine. Ram pumps were also used, as in the Williams' Shaft of Dolcoath, or today at Geevor.

It was again during the seventeenth century that the other major barrier to extensive mining was overcome—with the introduction of the art of 'shooting the rocks'. Metal tools had been developed which could bore holes for the slow wedging and lime breaking techniques, and now ways of using the more effective gunpowder were worked out. In the early 1600's at Chemnitz in Hungary, Gasper Weidl first used gunpowder for blasting. The method spread to Germany by 1627, and Prince Rupert, son of the Queen of Bohemia, who had interests in mines both in Germany and England, caused the technique to be brought to his English mines. Here the Goldophins again played an important part, for in 1689 they asked Thomas Epsley, who had learnt the way of using gunpowder in the English mines (those of the Mines Royal in Somersetshire), to come and show the art to the miners of the Wheal Vor district. It is probable that at about the same time the use of gunpowder started in mines of the St. Agnes district. Gunpowder soon became of major importance to Cornish mining. At first, as in lime breaking, gunpowder was used in shotholes of about 3in. diameter, but by 1720 it had been found that the more quickly drilled holes of about 1½ in. were sufficient. A problem, which took some time to resolve, was how to transfer the firing flame safely to the charge, and small 'pipes' of quills or straws were used at first.

The holes were drilled so that the powder could explode to a 'free face'—it was not until the advent of high explosives that holes drilled 'straight in' could be used (see fig. II). The powder was tamped in so that when it exploded, generating great heat and pressure, the rocks at the side of the hole were burst away. The loading and stemming had to be done carefully, and the hole was scraped clean and dry. If needed, a wet hole was sealed by ramming down a ball of clay with the conical 'claying bar', so driving clay into water bearing cracks—the powder being rapidly destroyed by water. Next the gunpowder charge judged to be correct was loaded into the bottom in a charging spoon (or wrapped up in a paper cartridge if the hole sloped upwards). A long tapering rod—the needle—was put to the powder, and clay stemming rammed in with a tamping-bar. The needle was carefully pulled out, and the quill or straw fuse pushed down the little tunnel left to the powder. This fuse was made by pushing the narrow end of a quill or straw into the wide end of the next, making a long, thin, tube. Some gunpowder was then 'bruised' to a powder on a dry shovel, and the top of the long tube held between the fingers in a cupped hand. The powder was poured into the hand and trickled down to fill the tube. When all was ready to fire, the end of a candle was fixed under the straw coming out of the shothole, so that the candle would burn for a short time after being lit, and let the miners beat a retreat. The candle was lit, the miners left, the powder flashed off down the tube, and the charge exploded.

Firing gunpowder in this manner was risky. This led William Bickford in 1831 to invent 'safety fuse', and build a factory for its manufacture in Tuckingmill. Bickford saw a ropemaker at work, and had the idea of letting gunpowder form the core of a rope, which would make an even burning fuse, and which could be covered with guttapercha and jute cord counterings to make it waterproof and strong. The first safety fuse was made by a man trickling gunpowder from a funnel strapped to his waist into the centre of a fuse rope of jute that he spun walking backwards down a 24 ft. 'walk'. This gave fuse in 24 ft. lengths—and to this day safety fuse is frequently sold in 24 ft. units. The fuse burnt regularly at 30 seconds a foot, rather than the unreliable flashing-off of powder down a straw tube. Later machines were designed to spin the fuse continuously. After spinning, the 'carcass' was treated by running through molten bitumen and guttapercha, and then further counterings of cord put on to give strength. A final dusting of china clay helped to show up the fuse where it emerged from a shothole. Only within the last few years has safety fuse manufacturing stopped in Cornwall—the fuse now being made in Scotland. The main factory was Bickford Smith's in

Tuckingmill, but numerous others, such as Bennets of Roskear, also made fuse.

Gunpowder was an important mining material. Factories were set up in Cornwall, for instance at Kennall Vale. The powder was made by first rough mixing the ingredients: 70%-75% saltpeter, 10%-15% charcoal, and 10% sulphur: and then milling extremely fine. The mills for doing this were made of two large stone wheels, weighing about 4 tons each, which were rolled round on their edges for up to an hour on the rough mixture put on the mill bed. Towards the close of the nineteenth century iron mill wheels were adopted. The 'mill cake' was then pressed into hard sheets, which were broken down into small granules, so that the explosive would ignite and burn evenly. Finally, the granules were rumbled in a barrel to glaze for easy loading, and sifted. The manufacture was dangerous, as a spark could cause an explosion. Like all explosive factories, the plant was spread out, so that any explosion could be localised. These factories were in quiet valleys, using waterwheels to drive the mills.

In the 1860's further advances were made to aid rock breaking. Nobel developed compositions which enabled the powerful and dangerous liquid, nitroglycerine, to be used in a form sufficiently safe to be applied in mining. Nitroglycerine is a high explosive—gunpowder burnt rapidly in a shothole, generating the pressure to burst out the rock, but nitroglycerine detonated—an exceedingly intense and rapid disintegration which produced great pressure and a shock wave which would shatter hard rock. Weight for weight high explosives were about three times as effective as gunpowder, thus smaller and fewer shotholes were required whilst the fumes released were not so noxious as those from gunpowder.

Other high explosives were also discovered at this time, but those based on nitroglycerine were particularly liked in Cornwall as they were extremely powerful and could be made into a water resistant jelly. Blasting gelatine (made by dissolving a small quantity of nitrocotton in warm nitroglycerine) was very popular, particularly in the 'tight' work of the St. Just mines. Another early high explosive much used, which was cheaper but not quite so powerful, was gelatine dynamite (thin blasting gelatine mixed with a 'dope' of potassium nitrate and woodmeal). Flame or sparks will not reliably initiate the detonation of high explosives, and a detonator, or 'cap', which provides a violent explosive shock, must be used. The cap is crimped on to the end of the safety fuse, and buried in a cartridge of the charge. At first the cap was a thin copper tube loaded with a mixture of fulminate of mercury and potassium chlorate; but since 1940, it has been made usually of aluminium composite charged with a high explosive such as tetryl, and a top load of a lead azide mixture to initiate the detonation.

Factories for high explosives were built in Cornwall to supply the mines; in 1888 the National Explosives Co. at Hayle Towans, and in 1892 the British and Colonial Explosives Company at Cligga Head. The Hayle factory was designed by Oscar Guttman, another famous name in the explosives world, and was the first in Britain to use the 'continental' process for the separation of the nitroglycerine from waste nitrating acid. The old British process was to skim off the nitroglycerine by hand; at Hayle the mixture was put into a closed vessel and the acid drawn off from below. A nitroglycerine plant is called a 'hill'—the installation often being built down the side of sloping ground. The glycerine was nitrated in a water-cooled lead tank, by mixed nitric and sulphuric acids at the top of the 'hill', then separated and washed as it flowed down from building to building. Each process building was thoroughly barricaded by heavy mounds, which can still be seen. After mixing, the explosive was cartridged by girls working in many small individual huts, again spread out to localise the effect of any accident. Large acid making and recovery plants were also part of the factory. Much of all this at Hayle is now a holiday caravan site—a change indeed. With the slump in Cornish mining in the 1920's the factories were closed down.

The boring of shotholes is an important part of mining. The early method was to use long rods—jumping bars—with a heavy bulge along their length to give weight. The cutting edge was forged to a square, later a chisel shape, slightly wider than the rod so as to prevent jamming and allow the rock chips to pass. The metal was relatively soft and required frequent re-sharpening. This jumper bar type of drill was employed by lifting and letting fall, turning at each drop to chip out a cylindrical hole. Using a mallet to beat a drill was quicker, and was normal rock drilling practice until well into the twentieth century. The drill rod, 'borer' or 'steel', on the other hand, was a plain rod, with the cutting edge forged as before. In single handed drilling (much used in the narrow workings of the mines of the Land's End peninsula), the steel was held and turned in one hand, and beaten with a hammer held in the other. The drill was frequently grasped with all the fingers and thumb curling one way on the steel—a missed blow was then slightly less agonising. One has only to try drilling a few inches in this way to appreciate the immense amount of labour and skill of the past miners. In multi-handed boring (particularly favoured in the Camborne mines), one man held and turned the steel, and usually two others beat the drill alternately. Some remarkably high speeds were reached in this manner, a good team drilling over an inch a minute in granite using one inch diameter steels. A set of steels was used with bits which were forged narrower and narrower as the hole increased in depth, to allow for wear and stop jamming. The final 'back' of the hole would be about 1½ in. diameter for 1¼ in. cartridges of explosive. A large mine used a great number of steels, which had to be re-forged and ground sharp at the end of the shaft. In some mines this was done underground, but mostly the steels were hoisted to grass at the end of the shift in special waggons. These rows of steels made a fine sight as they were trammed to the sharpening sheds.

Drilling machines were introduced in the 1870's. Fig. V shows an early example, the Barrow made by M. Loam and Son, which was mounted on a 'bar', screw clamped into the working place. Compressed air at from 40-80 lbs. per square inch was fed to a piston, which had the steel clamped on by bolts. A mechanical valve operated by the piston controlled the air feed to give the power stroke, and light return stroke. The steel was rotated by gearing off the crank used to screw the machine forwards. Later designs adopted air operated valves, which were more adaptable. The drill steel was 'clouted' with a hammer to prevent it from jamming, and it was soon found that water had to be sprayed into the hole to cut down dust. The silica drill dust was inhaled by the miners and caused the lung tissue to become fibrotic and inactive—causing silicosis. To be more effective, the steels for machine drilling were forged with a cross-bit cutting edge (two chisel edges at right angles). Drilling machines are now most important pieces of mining equipment, and have developed into wonderful pieces of precision engineering. Water is fed through the centre of the machine and down a hole cast in the drill to the cutting edge, preventing dust formation. The speed of working is high with some machines—approximately 3,000 blows per minute, drilling up to a foot of 1½ in. hole a minute in granite. The steel being turned at 200 r.p.m. through a rachet and rifle bar, gives the appearance that the drill cuts by rotation, the steel being struck by the piston instead of clamped to it. Since 1945, the drill steel itself has been improved, an insert of very hard tungsten carbide being used as the cutting edge, so that many holes can be completed without a change.

Large air compressors were installed to supply the machines, the compressors being normally built near the pumping engine or the hoist, so as to have the boilers centralised. The early designs used a pair of pistons on a horizontal rod, connected to a flywheel, one working in a power cylinder fed with steam, the other in the air compression cylinder tandem fashion. Later, two of these pairs were coupled, one each side, to the flywheel, to give smooth running, when the steam was used

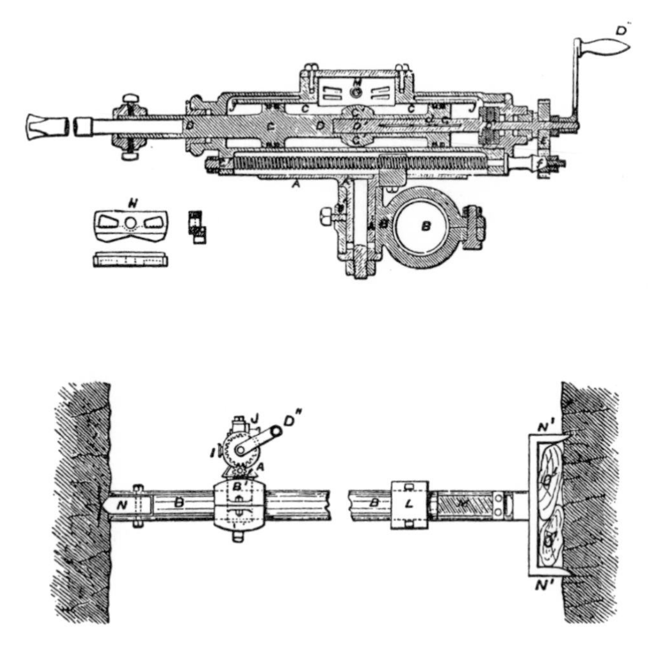

FIG. V SECTION (ABOVE) AND END VIEW (BELOW) OF THE 'BARROW' ROCK DRILL

cross-compound. The air was compressed to about 60 lbs. per square inch in the early single stage compressors, 80 lbs. per square inch in the later two stage machines, and stored in a steel tank 'receiver'. Dolcoath mine installed a particularly fine example of this type in 1902, of 2,000 cu. ft. min. capacity, and which used an average of 100 tons of coal a month. Today, Cornish mines use compressors with two vertical cylinders. The air is first taken to about 40 lb./in^2 in a large diameter low pressure stage, and then to 80 lb/in^2 by the smaller high pressure stage. A typical installation of this type, at Geevor, supplies up to 1,500 cu. ft. per minute, and is powered by a 300 b.h.p. electric motor.

The methods of mining developed with the introduction of these machines and aids. New approaches to problems could be devised which overcame age old difficulties. The miners were relieved of much of the drudgery of early mining, and were able to explore more extensively.

A PAIR OF HORIZONTAL AIR COMPRESSORS AS USED IN CORNISH MINING C. 1880
(reproduced from a contemporary engraving in the catalogue of Harvey & Co., of Hayle)

CHAPTER FOUR

UNDERGROUND MINING

UNDERGROUND mining methods were evolved when the original trenching or coffin workings into a lode from surface began to reach depths which required an elaboration from simple pitting to 'underhand' working in order to reach the ore. Thus there was some overlap between true surface working and underground mining, depending on the nature of the ground. To expose the ore in depth, horizontal tunnels, or 'levels', were driven into the lode, and methods of working upwards into the lode from the roof ('back') of a level grew into 'overhand' mining methods. Going below the surface or 'grass', the working place in the lode became a 'stope' and the winning of ore 'stoping'. When stoping in a block of ore was completed, the cavity left was termed a 'gunnis'.

During the sixteenth century the use of adits for draining the water became more widespread and miners from the Harz district were brought to Cornwall to employ their experience. Underground mining activities steadily increased from that time. Although it is probable that shaft working had been practised prior to this, the adit controlled the layout of the mine because of the importance of the water drainage. Shafts were used in a subsidiary role, for ventilation and perhaps some hoisting, but it was not until the introduction of the steam driven pump that deep shaft mining could be undertaken, after a transition stage when water-wheels and horse driven pumps had enabled some work at shallow depth.

Underhand and overhand stoping methods formed the basis of the underground winning of ore in Cornish mines. At first the lodes were roughly worked from the floor or back of a level, but more systematic approaches were applied as experience was gained. After prospecting shafts and levels had been opened, the lode was explored and proved by 'blocking out' the ore into sections by further levels and small intermediate shafts. The amount of this 'development' varied. In some mines the "eyes" were picked out in a haphazard fashion, leaving much good ore unworked. In others, the lodes were examined and developed so that the captains knew in advance which would provide good values, and so realised full return for the work. The main principles of mining have remained unchanged, but there have been continuous refinements in techniques. The choice of method was controlled by the position of the values, and in some instances, the characteristics of the ore.

As far as possible, the stoping was arranged so that the broken ore would fall into positions where it could be conveniently taken out of

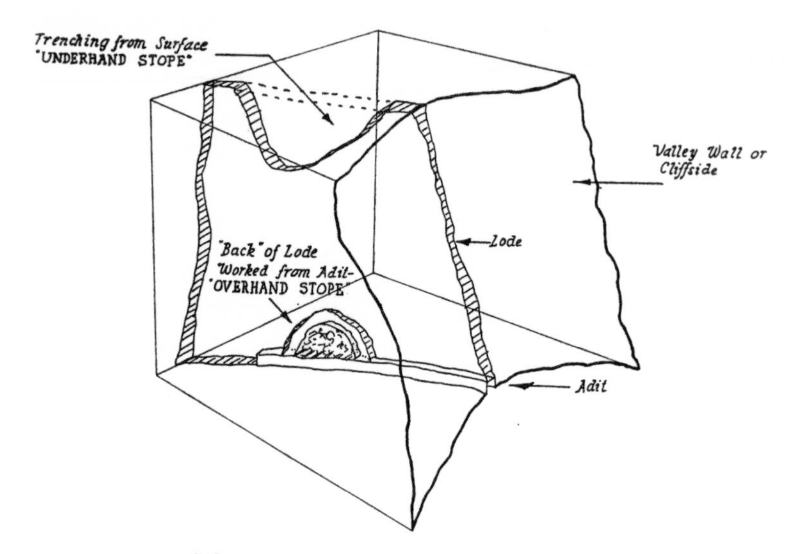

FIG. VI ISOMETRIC VIEW OF LODE AND STOPES

the mine by adit or shaft. This reduced the amount of handling, and was helped by the dip of the lodes almost always being such that the material could be directed to slide to a loading place. More levels to open up the ground and handle the ore were driven into the lode at intervals as the workings went deeper, usually every 10 to 12 fathoms (up to 20 fathoms after the nineteenth century). In early underhand stoping the miners dug into the lode in the floor of a level, the broken ground being drawn out in a kibble (a barrel-shaped bucket of wood or iron) with a windlass, and loaded into wheelbarrows which ran on planks or, during the nineteenth century, into small waggons running on rails in the level. In some cases, the ore was cast up (shammelled) from step to step into the level. Later, small intermediate shafts (winzes) between levels were put down at horizontal intervals of about 50 fathoms. The stoping then

CORNISH PUMPING ENGINE. The 90-inch New Cooks engine at South Crofty, 1924. The cylinder cover, with piston rod going up to the parallel motion and bob above is in the foreground with the nozzles behind.

STEAM HOIST. The steam hoist at Robinson's Shaft, South Crofty, built by Holman Bros., of Camborne in 1907. This photograph was taken in 1965 and an electric winder has now been installed in its place.

TIN DRESSING FLOORS. General view of the dressing floors at Wheal Grenville, near Camborne, taken in 1904. The Cornish stamps with their driving engine, and also circular concave and convex buddles can be seen. This photograph gives an impression of the general appearance of dressing floors in the nineteenth century.

PRELIMINARY 'GREEN' CONCENTRATION OF TIN ORE PULP being made at Dolcoath on Frue vanners in 1904. The belts of the vanners are running up towards the centre of the picture, being given a side shake as they move round. The tin concentrate can be seen coming over in light streaks, which will be dropped off into tubs below the vanners.

commenced at the top of the winze, the ore falling down it for hand loading into waggons, or into a timber chute (or 'mill') at the bottom which could be discharged into trams. With the introduction of explosives, the ore was blasted in a series of steps or benches, working down and outwards. Benching in this manner facilitated the drilling of shotholes. and the fragmentation of the ore. The benches were carried at an angle which would allow the miners to climb up and down, but sufficiently steep to permit the broken ore to be shovelled and worked down to the lowest point. Fig. VI shows an underhand stope worked between the top and middle levels. The upper level was kept open as a travelling way by making a floor of timbering, hitched into the sides from hangingwall to footwall, where the original floor had been worked out.

In overhand or 'back' stoping, the mining commenced by working upwards from the back of the level. Timbers could then be hitched into the sides, making a protective cover to the level, and a platform for the ore to land on. Such timber floors and roofs in underhand and overhand stoping are known as 'stulls'. In the early overhead stopes the ore was allowed to fall through gaps in the stulls, and was then shovelled into waggons, but this required a considerable amount of double-handling. Subsequently, 'mills' or chutes were built into the timbers, spaced along the level. The miners blasted the lode in benches, up and outwards, standing on the broken rock. As the broken material was more bulky than when *in situ*, some was drawn out of the stope through the mills as the stoping progressed, allowing the miners room to work and at the same time taking ore to 'grass' (surface).

In copper mining during the eighteenth and nineteenth centuries, stulls were also built into the stopes as platforms for waste. The ore was hand picked in the stopes so that the stuff sent to grass was as rich as possible, the gangue being packed on the stulls. With some tin ore, it was difficult to use hand picking as the cassiterite was frequently locked up in the waste as small crystals and could not be distinguished. Instead the lode was sampled to determine the width worth mining for the economic concentration of values, and the miners then stoped the lode to this 'assay width'.

Using these techniques, the levels progressively became transformed to artificial floors and roofs of timber, which could weaken, and had to be maintained. More recent mining methods developed the levels in the footwall rock to the side of the lode, so that the tracks could run on solid ground which would not be removed. Nowadays 'stope drives' or 'interdrives' (introduced in the 1890's) are also made, which are small levels carried just under or over the main level. Connections ('boxholes') are made into the main level at intervals, to allow the miners access to the stope face and provide a way for compressed air and water pipes, and to act as channels leading from the bottom of the stope to the mills. The untouched lode between the stope drive and the level replaces the stull timbers, and acts as a pillar of solid supporting ground. In some parts of a lode, the ground might be low in values, and not worth mining. These poor places were left as pillars. With the changing prices of tin, it was sometimes worth mining into these old pillars and recovering the ore.

The skill of placing and drilling shotholes was an important feature of stoping. Where possible, hand drilled holes were made either horizontally, or to slope downwards. Hand boring a hole upwards was difficult, as the power stroke on the hammer was 'up' and required a long underhand swing. Gunpowder loading was straightforward, but with high explosives it was found that the best results were obtained by spacing the charge out, the shock energy then coupling into the rock at intervals and chopping the lode up to give readily handled blocks of ore. Fig. VII shows the application of blasting gelatine in the mines of the Land's End area during the late nineteenth century. Each cartridge was slit into four lengths, the hole being charged with a length of gelatine, then a handful of fine rock until sufficient was loaded. Blasting gelatine was a sensitive explosive, and the shock from one slice detonating after traversing the fines was adequate to initiate the next. One slit length of gelatine was retained. This was lit from the candle, and burnt with a very hot pointed flame, which was used to light or 'spit' the fuse, the gunpowder fuse-core having been exposed by cutting the end at an angle. This slitting up of explosive and charging with dirt spacing would now be considered dangerous practice. Today, the charges are spaced out in a similar manner by using sticks of wood, cut about 9 inches long by ½ in. square. These sticks are loaded between the cartridges, the miner breaking one in half if he considers a more concentrated charge is required.

By the 1880's, air powered drills were being increasingly adopted by the mines. Jackhammers were designed for underhand stoping—light hand-held machines, which would quickly put down the 2 to 6 ft. holes used. 'Stoper' machines were also developed for overhand drilling. This was similar to a jackhammer, but had an air-operated ram working in a cylinder attached. This air ram pushed the machine up when a thumb-operated valve was pressed, working against the floor of broken ore, so that vertical holes could be bored. In this manner, the overhand stope was drilled in benches, using holes 5 to 8 ft. deep. Stopers are easy to use, the air cylinders taking the weight and forcing the steel into the rock.

Blasting is usually done at the end of a shift, and time allowed for the fumes to be cleared by the air currents. The miners cut their fuses to give

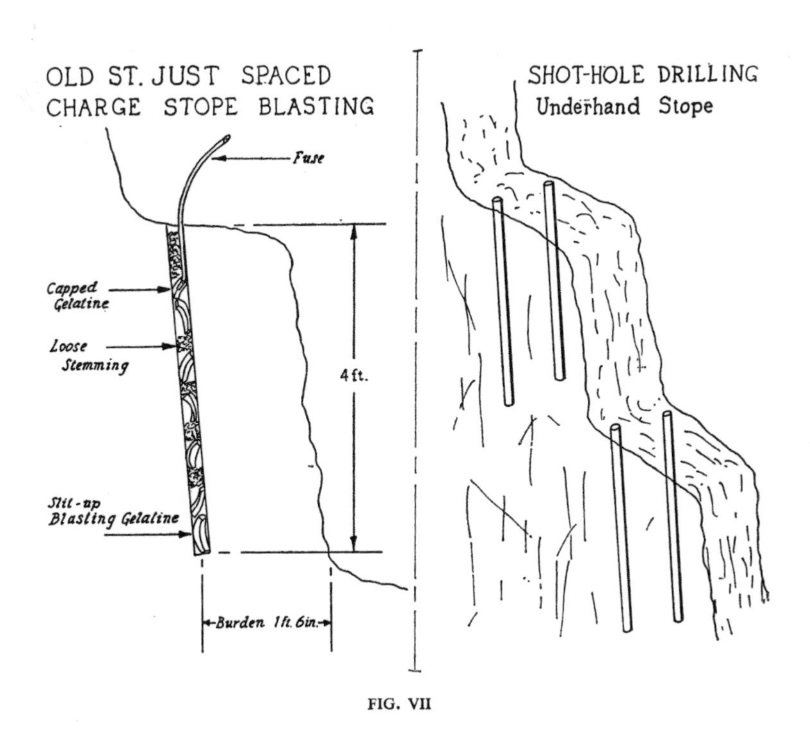

FIG. VII

a safe delay, and nowadays use the hot flame of an acetylene lamp to light-up. The timing of the lighting of the fuses is such that each shot will fire in sequence, and not break into a shothole out of rotation, cutting the explosive column and fuse. The miners ensure that they are well away along the levels before blasting starts, and count the shots as they detonate to check that all have fired. With the complicated layout of most mine workings, particular care is taken when shotfiring to see that no-one can walk into a dangerous place.

Blasting time is impressive underground. Each shot makes two reports—a 'knock' as the shock from the detonation reaches one through the rock, followed by a reverberating 'boom' from the blast carried in the air, the velocity of sound in rock being faster than in air. Many 'knock-booms' go off, and the air shivers with the blasts, sometimes shaking out the flame of one's acetylene lamp. Then the shots are over, and everything is quiet, waiting for the fumes to clear so that the next shift can work away the ore and prepare for hoisting.

The earliest lighting was by the weak tallow or oil dips, which were displaced by candles. These were carried in bundles tied together with their wicks, and were used fixed into clay which could be dabbed on to the resin-impregnated felt helmets, or on to a handy hammer shaft or ladder. Candles were an important item of trade for the merchants supplying the mines until the 1900's, when acetylene lamps came into use. Today, electric caplamps are being introduced, and the cheerful rattle in the miner's pocket of a little tin of carbide for his lamp is becoming a memory of the past.

The miners used to be paid as 'tut-workers' or 'tributers'. Tutwork was a system of piecework, the miners taking a contract and being paid a set amount per unit of work done. Thus payments could be per fathom advance in a level of given size (usually 6 ft. by 4 ft.), or per 'square fathom' of lode stoped (actually per square fathom advance over the lode width). Alternatively, the unit could be weight mined. The mine owners provided the necessary materials—tools, candles, dynamite—and the value of these was deducted at the end of the contract. In 'working on tribute', a miner, or 'pare' (gang of men) was paid a proportion of the value of ore won. The tributing miners bid against each other for the 'pitch' they would work on 'setting' days. The miners gathered outside the mine office (count house) and the manager read out the working places available for tributing, asking for bids for the contract on the proportion of the value to be retained by the men. Sometimes the tribute was fixed more privately between men and manager direct.

The system encouraged skill in the miners' ability to judge lode values and signs of richness or poverty as well as helping the management, as men were always on the lookout for good ore. The poorer the pitch, usually the higher the price bid, so that if an unexpectedly good bunch of ore was struck, the miners developed many ways of preventing the mine captains—who had often been tributers themselves and fully aware the men were likely to try and outwit them—from learning the fact, the ore being carefully concealed. Private agreements were frequently made between tributers working pitches of rich and poor ore, the pare on low tribute being prepared to transfer part of their ‘pile’ secretly, for a consideration, to a pare with a high tribute price, to the benefit of both parties. A tribute might be fixed at ‘1/5’ or 4s. for every £1 realised by the ore, the cost of materials supplied by the mine and all expenses of hoisting or ‘drawing’, as well as sampling and dressing, being deducted. This system encouraged the men to work well, but also tended to reduce the supervision given by the captains with the result that the miners might become careless and dangerous practices develop unchecked. Tributing also could, as one writer expressed it, “embolden the miners to duplicity, with evil results to both the mine and the neighbourhood”. Although the tributing system was often used in tin mines, particularly when the mine was old and experienced miners could pick out what patches of rich ore remained, it was not as valuable as in copper mining. Tributers might take out the rich bands in a tin lode, leaving the lower grades behind in the stopes. Although this resulted in a high grade ore being sent to the dressing floors, much good lode-stuff which could have been profitably treated was wasted. The overall economics favoured the leaving of low values in copper mines, but not in tin. Some men were employed as ‘labourers’, usually on surface work, and were paid a fixed wage. Today the terms tributer and tutworker have passed into history and forms of piecework and daywork payment are in use instead.

The country rock making up the hangingwall and footwall of most Cornish mines stood well, and did not require heavy propping and supporting. Pillars, or sections of lode, were left where the hangingwall did not ring clearly when sounded with a hammer, or where it looked unsafe, and these were normally sufficient to prevent collapse. In some mines which had lodes that were wide and flat-dipping, and where the hanging wall was likely to be weak, timbering had to be used. Very heavy support was needed in some of the Camborne mines where there could be falls of ground—for example, in the deep workings on the Dolcoath Main Lode.

The waggons of ore from the stopes could be treated in several different ways. They might be trammed along the level to a shaft for hoisting by kibble or skip, or later to an ‘ore pass’, a small subsidiary interconnecting way down which the ore could be tipped from the various working levels. This pass led to a loading pocket at one of the lowest levels. Sometimes an old gunnis was made use of as part of the pass.

The ore was tipped from the trams on to a coarse sieve made of old rails, known as a 'grizzly', leading to the pass. This held back large stones which might cause blockages. The ore through the various grizzlies fell to the main pocket, hoisting being from this one station only, simplifying working. The big blocks held on the grizzly were broken by 'sandblasting' with explosive. Sufficient dynamite was put on the block, covered with mud and fines, and fired. As the explosive was practically in the open, it was the sheer violence of the shock of detonation that shattered the stone. Only high explosives could be used in this way. Although more explosive had to be used than would be needed in a shothole, it was usually cheaper to sandblast than to keep expensive drills standing by for such relatively unproductive work.

Because of the numerous workings essential to develop the lodes, and handle broken ore, a plan or section of most Cornish mines at first sight appears to be a bewildering maze of levels and shafts. Added to this there may be complications of the lodes, which cause the levels to deviate and branch. Moreover, several lodes running parallel to each other may all be stoped in various places by one mine. The distances between the levels and shafts depended on the working conditions—what could be handled in the width of ore worked, how the valuable 'shoots' were placed within the lodes, and the detailed way in which the miners worked. Much development was needed to allow the orebody to be explored and sampled to determine what would be worth working at various market prices of the metal. As the prices changed with economic conditions, so parts of a lode could be considered payable, and worked at a profit. Added to 'on lode' development, levels were put out at right angles to the main direction of work, to explore for fresh parallel lodes, although many of the old Cornish mines neglected to do such crosscutting which could have revealed valuable deposits. These 'crosscuts' were developed when geological conditions indicated favourable ground for prospecting. Nowadays diamond drills are used to aid in such exploration—being more economical than mining an expensive crosscut in many cases. The drill is made up of diamonds set around the end of a pipe-like drill head. This bit is rotated and fed into the rock, cutting out a core. The bit is then removed and the core contained in it extracted and examined for values. This is repeated so that a long series of cores can be logged, cutting through much ground.

In the past, development and stoping tended to be more haphazard than today. If a bunch of good ore was found while driving a level, it was frequently worked into without being blocked out. This was particularly the case in mines which were in some legal dispute such as over the rights as to who owned the ore, or in mines controlled by financiers wanting quick returns on investment. The workings of Wheal Vor are a classic case of a mine which was stripped of its rich ore, regardless of future working.

Before the introduction of gunpowder, the driving of levels by

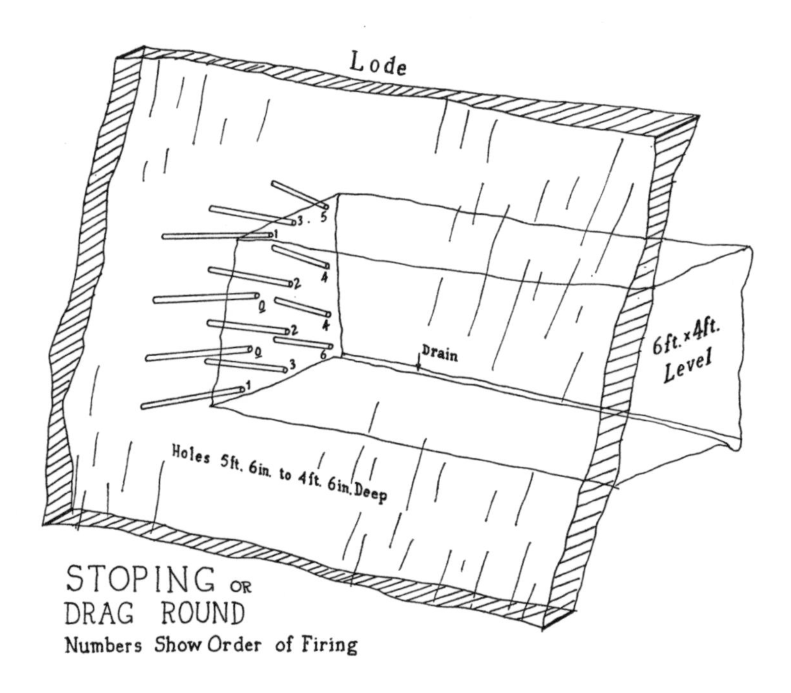

FIG. VIII

'hammer and gad' or by fire-setting to open up the ground was most laborious work, and hence development was restricted. Gunpowder was thus an important factor in the expansion of underground mining. Hand drilled holes were bored so that the explosive could break the rock out section by section, to a 'free face'. Wherever possible, a natural weakness was used to help the gunpowder do its work. A lode

OVERLEAF: ISOMETRIC VIEW OF THE UNDERGROUND WORKINGS AND SURFACE LAYOUT OF A TYPICAL CORNISH TIN MINE, C. 1905

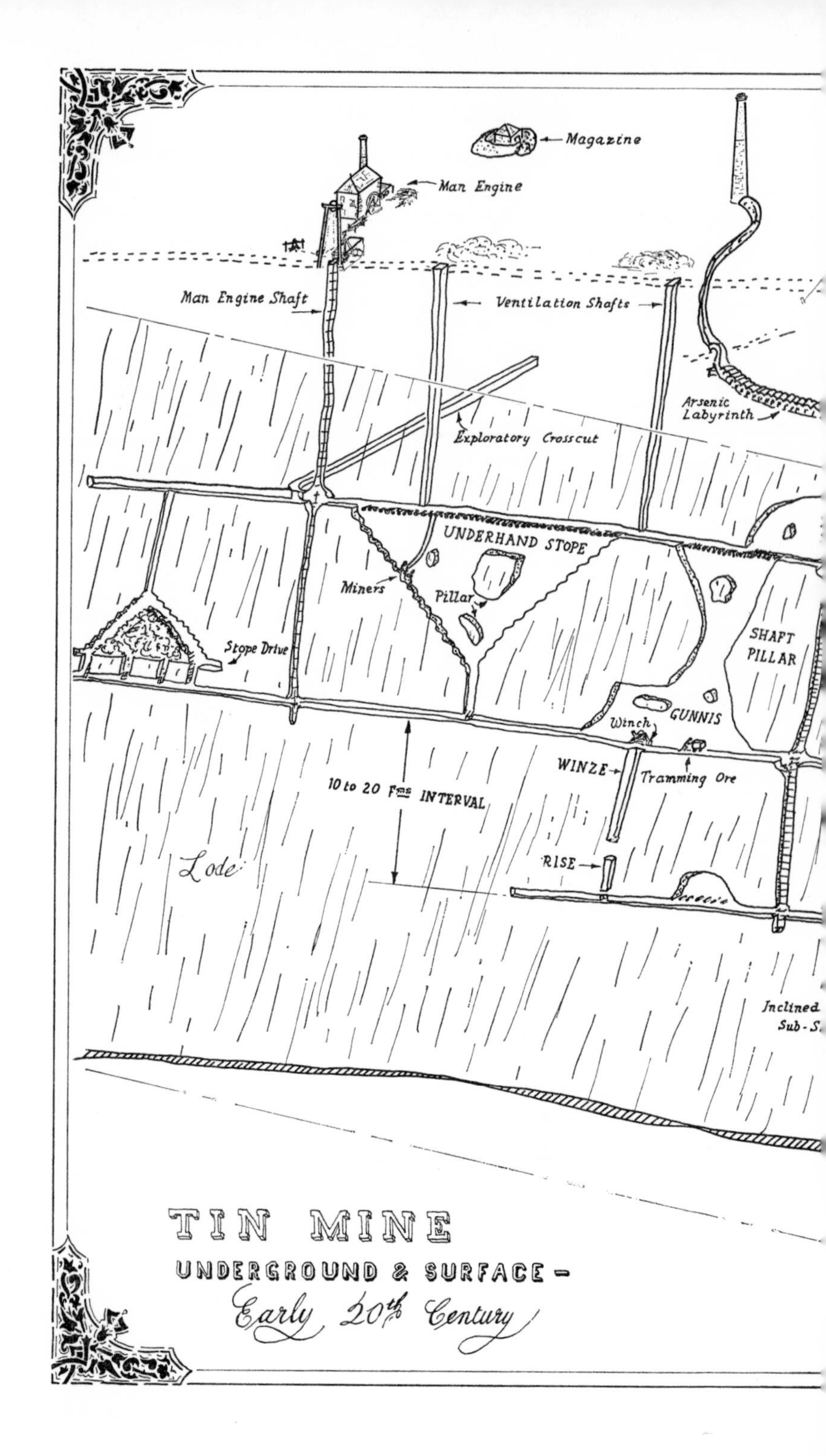

Magazine
Man Engine
Man Engine Shaft
Ventilation Shafts
Arsenic Labyrinth
Exploratory Crosscut
UNDERHAND STOPE
Miners
Pillar
Stope Drive
SHAFT PILLAR
GUNNIS
Winch
WINZE
Tramming Ore
10 to 20 Fms INTERVAL
Lode
RISE
Inclined
Sub-S
TIN MINE
UNDERGROUND & SURFACE –
Early 20th Century

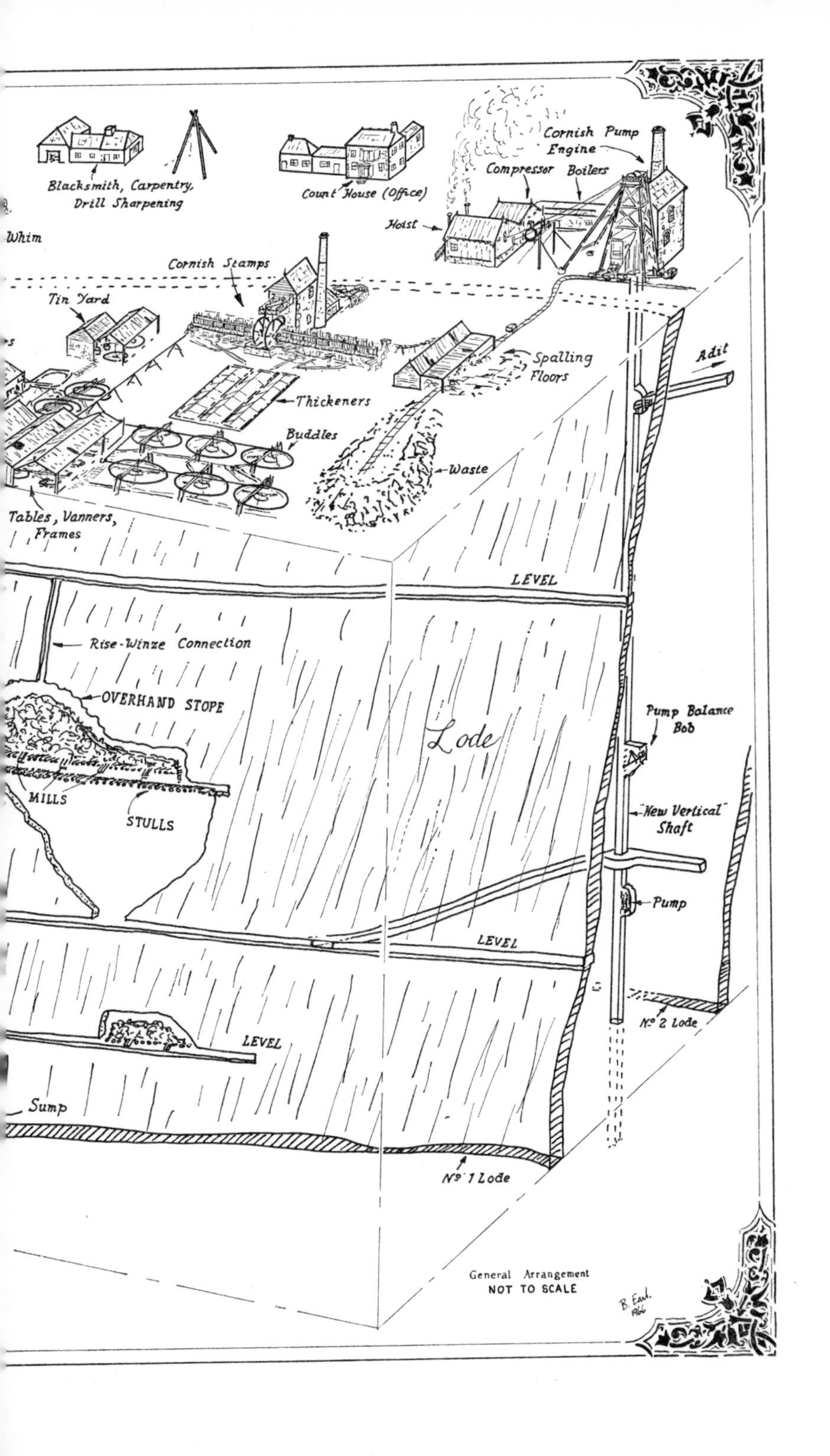
Blacksmith, Carpentry, Drill Sharpening
Count House (Office)
Cornish Pump Engine
Compressor
Boilers
Hoist
Whim
Cornish Stamps
Tin Yard
Spalling Floors
Adit
Thickeners
Buddles
Waste
Tables, Vanners, Frames
LEVEL
Rise-Winze Connection
OVERHAND STOPE
Lode
Pump Balance Bob
MILLS
STULLS
"New Vertical" Shaft
Pump
LEVEL
No 2 Lode
LEVEL
Sump
No 1 Lode
General Arrangement
NOT TO SCALE
B. Earl. 1966

tends to provide such a weakness in the rocks, and so the levels were driven 'on lode' whenever practicable. The lode could be picked away in a soft plane such as a clay slip, to give an easy path for the blast. In the St. Just mines, the tool used for this was called a 'peeker'. The shotholes were angled in (see Fig. VIII), the first holes to be fired being placed so as to allow the energy to reach both the weak lode plane and the face of the level itself. The following shots then broke into the cavity produced by the earlier charge, and so enlarged the level to its final size—approximately 6 ft. high by 4 ft. wide. An advance of 2 to 3 ft. per 'round' of holes was considered good. In crosscutting through granite country rock, the planes of weakness in the granite itself were utilised. The mining was slower and required even more skill than on-lode work, to take advantage of these weakness planes. In the 1870's heavy bar-mounted compressed air machines were introduced for drilling in development work, the first use of machine drilling in Cornish mining.

With the introduction of high explosives, great improvements were made in driving levels, shafts and crosscuts. The shattering effect of the shockwave on the hard killas and granites enabled the holes to be drilled 'tighter', with less free face. The Pyramid cut (4 holes drilled to meet at a point and burst out a pyramid section of the face, later holes blasting into this cavity or 'cut') was developed, but the angling necessary for the cut holes imposed a limit to the depth that could be drilled—the machines meeting the side of the level. This led to the 'Cornish Cut' being devised (sometimes called a Burn Cut)—a series of closely spaced holes (on about 3 in. centres) drilled straight in along the axis of the level. Some of these holes were left open and uncharged. The gelatine was loaded in with wood spacers, and on firing the shock wave could act on to the uncharged holes, and allow the cut area to be smashed to small fragments. If no spacers were used the rock could fracture and then 're-cement' to a hard block so that the following 'easer' holes could not break out, and the cut fail. Until the 1960's, and the spread of electric delay firing, with which the whole round is fired complete, the cut and easer holes were normally drilled and fired alone, to ensure that the cut had pulled to the back. A Cornish Cut may be up to 7 ft. 6 in. deep, giving that advance to the tunnel per 'round' of holes. The remaining shotholes to enlarge the cut void were then drilled and fired. These enlarged the cut hole by hole in sequence to the full size of the level, and were given names—'knee holes', 'shoulder holes', 'side holes', 'back holes', 'toe holes' and 'lifters'—to indicate their position. The miners cut and cap their fuses to time the shots to fire in the desired order (also lighting up in the required order to make doubly sure), the lifter holes on the floor having the longest fuses and lit last. This gives a well fragmented rock pile, which is easy to load out. When firing a cut in a long development level, it is usually only possible to retreat a few

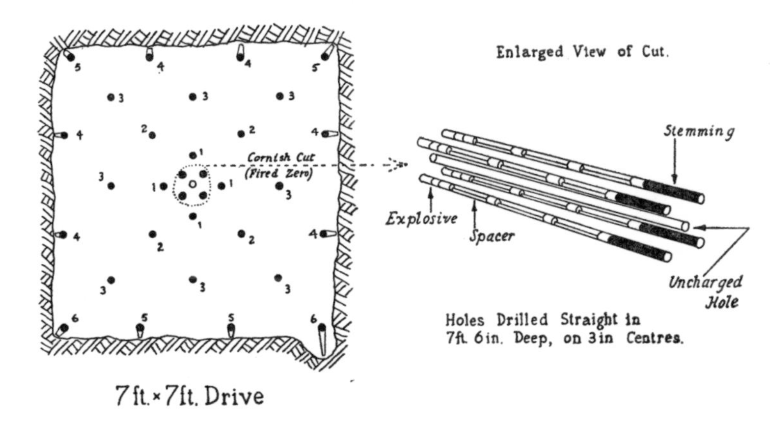

FIG. IX

hundred feet from the face to shelter in an upturned waggon. The cut shot, with a heavy charge of explosive detonating simultaneously, goes off with a mighty bang. When this is first experienced, it gives the feeling of being like a bullet in a gunbarrel.

In present methods of driving, the services used have to be kept up to the face, such as pipes for compressed air and water, and also metal and canvas ducting for fresh air from a booster fan working in a clean air section. The miners also lay the track for the waggons to take out the broken rock as the level advances. In the past, the rock was loaded out by hand, but now a rocker shovel loader is sometimes used. This comprises a shaped bucket pivoted to a compressed air-operated chassis, which is driven into the rockpile. The bucket is then lifted up and over the chassis into a waggon coupled behind. Light jackhammer drills on airlegs are popular now for drilling the shotholes. The airleg is similar to the air cylinder on a stoper, but pivoted so that it can push the jackhammer horizontally, the airleg sloping to the floor. However, heavy 'drifter' machines which may be mounted on a hydraulic positioning carriage are also in use.

Because of the ease of driving with modern equipment, and the advantage of space for travelling and ventilation, a typical dimension of a level is now 7 ft. by 7 ft. Wherever possible, the waste rock, or 'attle', has always been trammed and tipped into an old gunnis. Hoisting barren rock ('deads') to the surface is avoided as time-consuming and wasteful of hoist power. Small battery electric locomotives are used for tramming. Also the round is frequently blasted complete in one operation, using electric delay action detonators in place of cap and fuse. These have a range of set delay times made up of an even burning tube of composition pressed on to the base charge which is ignited by the flash from an electric fusehead, triggered by an 'exploder' from a safe point.

The winzes—which may be formed as 'rise-winze connections'—have usually been mined in two sections. The lower part is 'risen' by blasting upwards from the lower level, and the upper sunk or 'winzed' from above. Accurate surveying makes sure that the rise and winze hole together and connect. Both parts are made about 4 ft. by 6 ft. To start shotholes are drilled from the side of the level in the lode, the drilling pattern being similar to that used in driving, but "stood on end". As the rise section goes up, holes are bored into the sides, and old steels put in to form a base for wood board platforms, which the miners can work from. In the past the difficult 'upper' holes had to be drilled by hand, but now stopers working off the platform are used. The material from the blasts falls to the bottom and is 'mucked out' into trams. Reaching the top of a rise, after clambering up chain ladders to the platform on old drill steels, one meets a curious form of controlled pandemonium. The stoper roars within inches of one's ear, raining blows on the drill. Misty, oily air pours out of its exhaust. It is hot and cramped, with miners, drills, shovels, hammers, air and water pipes. Often a stream of hot water sprays out of the rocks. The rock face has to be watched carefully, making sure the drilling is not shaking loose any stones which could fall.

In winzing, small kibbles with a windlass or compressed air hoist are used to bring out the broken rock. Nowadays drilling is by jackhammer. Mucking out a winze can be difficult because of the confined space, and having to dig down into the rockpile.

Shafts were normally larger than a winze (although the earliest of the former were only about 3 ft. by 4 ft.) and extended through the workings, having in addition some form of timbering to carry hoisting, pumping and other services for the mine, and for support. As simple adit and coffin mining spread, shafts gave access to the deeper workings, and also provided airways for ventilation. Extra ventilation shafts were frequently sunk (along the strike at about 50 fathom intervals) as the mine extended—and the old workings on a lode could often be traced on the surface by the row of 'deads' from these shafts in heaps across the fields. This enabled a natural circulation of air to pass through the workings, which 'downcasted' into the mine through some shafts and emerged out of 'upcasting' shafts—in cold weather often wreathing the shaftgear in mist as the water vapour gathered by the warm air in the mine condensed. The circulation was achieved mainly by the heating of the subterranean air by the rocks. In the deeper workings, fewer ventilation shafts were sunk, doors sometimes being built into a level to guide the air flow. In difficult cases a 'duck machine' was used to ventilate an 'end', a wooden container moved in and out of another, sealed with water and equipped with flap valves leading to a wood pipe, the motion being imparted by coupling it to the main pump rod. According to the arrangement of the valves, air could be either drawn out, or forced into the working, the machine being the fore-runner of the present electrically driven fans coupled into ducting. The original role of the winze, which at times was known as a 'winds', was as an airway and only later was it found useful as part of the more general exploration and development work for stoping.

Many of the early shafts were sunk in the lode itself—'on lode'—which resulted in their being crooked—following the variations of underlie and twists of the ore. Alternatively when the position of the lode below was thought to be certain, a vertical shaft was put down to cut the lode at a suitable point, and then followed the underlie. In many cases the lode was worked away on either side of the shaft, so that it became little more than a timbered passage for kibbles, ladders, pump rods and pipes running through a gunnis. With better practice, some ore was left standing on each side to act as a pillar and prevent crushing. In olden times, the mines started from exploratory pits, and the shafts were kept small to make sinking more easy, but which later restricted their hoisting capacity. Timber lagging was put in to help the kibbles in their passage through the twisting shaft. As support in loose ground wood frames were blocked in, held at intervals on bearers—heavy timbers projecting into holes hitched 2 ft. in on each side of the shaft. In hard rock the frames were often dispensed with, cross pieces ('studdles') wedged against the sides being used where necessary to hold lagging and pipes. It is remarkable how under these conditions the 'old men' devised methods of working the pumps and rods as a mine grew larger and deeper, with such cramped and twisted shafts. An extreme example was the main shaft of Wheal Owles near St. Just, which had twelve pronounced changes of direction.

Until the nineteenth century, the hoisting was by rounded kibbles, so shaped to help them to slide past projections. The kibble was lowered to a level, and swung in to a wide section, the 'plat', where it was loaded. In many cases two kibbles were used—one being hoisted, the other uncoupled from the rope in the plat, to avoid loss of time from the slow shovelling and hoisting. The latter was effected by round or flat hemp rope until about 1820, when chains were introduced which packed well on narrow diameter windlass drums, and were flexible in the crooked shafts. They were, however, unreliable as a snapped link resulted in the kibble, or later the skip, crashing through the shaft. There were several accidents when men were hoisted by chain in the early gigs. Steel ropes began to replace chains after 1850, being much safer as a broken strand did not result in complete failure. Even so, chains were still in use for some work in the 1900's. With the larger outputs of the middle nineteenth century, 'skips' holding up to 1½ tons of ore were introduced into Cornish mines from the 1850's onwards. These were sheet iron, later steel, containers which ran in wood guides fixed in the shaft. Usually there were two wheels or U-shaped slippers on each side; very crooked shafts were equipped with skips having four wheels on each side, arranged so that on the reversed part of a curve the load was still carried on the wheels. The hoisting rope was retained in the central position in the shaft by numerous iron rollers on the concave side of the course. In many cases, the shafts were inclined, following either the underlie or strike of a lode, and this assisted the run of the skip. The skip was lowered to a plat, where a man could run the ore into it through a chute, the skip frequently being held by a swing-out block to prevent it from rolling down-shaft. When full and taken to surface, a 'lander' knocked a retaining pin out of a bottom dumping door of the container, running the ore out, hinged doors over the shaft preventing material dropping down. Later, the skips were loaded at the central loading pocket at a deep point in the shaft, fed from the pass. Originally, a single skip per shaft was used, but soon a second was added, running opposite so as to balance out the skip weight and increase the output. Skips are widely used nowadays, some having a pivoted, self-dumping container, and arrangements can be made (as at Geevor) to uncouple them and put on a cage to hoist men at suitable times in the shifts.

By the nineteenth century, the mines were employing large numbers of men—several had over a thousand—and reached depths of 1,000 ft.

or more. The men had to enter and leave the mine by ladders placed in the shafts, which was most time-consuming and damaging to health. To overcome these problems, a 'man-engine' devised in the Harz mines, was installed at Tresavean, near Lanner, in 1842. Platforms were put in

LOADING SKIPS IN THE SHAFT, WHEAL BULLER (REDRUTH)

a shaft every 10 to 12 ft., which was equipped with a wooden 'rod', similar to that of a Cornish pump, working with a stroke equal to the platform spacing. Hand and footholds were put on the rod, and the miners were lowered or raised by stepping from a platform on to the rod, being carried up or down a stroke, and then stepping off on to the next platform, and so on. Several of these man-engines were soon built, the power being provided by a waterwheel or steam engine. One early installation was at Fowey Consols in 1851, carrying the men 1,680 ft. vertically. The drive was from an overshot waterwheel 30 ft. by 6 ft. face, making 3 r.p.m. A crank on the axle of the wheel was directly connected to a bob, by which the rod was given a stroke of 12 ft. The rod was 8 in. square, built of timbers, 36 ft. long, butting together and connected by iron fishplates ('strapping plates') 12 ft. long. The platforms were 12 in. square on the rod, with 2 ft. iron bar handles. The weight was counterbalanced by three balance 'bobs', two of which were underground. At the end of the 1860's gigs (similar to skips but carrying men) were introduced, which could be hoisted in a shaft. These were particularly valuable in deep sections of a mine. However, a man-engine was still in use at Levant in 1919, when the top plate or 'main cap' broke causing the rods to collapse in the shaft, killing 31 men.

The earliest hoisting was accomplished by winding the rope on to a drum or 'whim', by manpower or horse, (in the very old vertical shafts the miners were sometimes brought up by standing in a loop of a rope, and this was hauled up by a windlass worked by two men). The horse whim could handle greater weights, and typically used two horses working round a 36 ft. diameter path, rotating a wood board drum or 'cage' about 12 ft. diameter by 4 ft. 6 in. high. The vertical axle of the drum ran in an upper bearing carried by timber framing, which also had the pulley for the rope or chain over the shaft mouth, the footstep bearing being a block of granite. The load was raised in kibbles holding about 2½ cwt. at 75 to 100 ft. per minute. The Cornish water whim was more complex. An overshot waterwheel was coupled by spur gearing to a sliding shaft on which were fixed two bevel wheels to gear with a third bevel connected to the winding drum. By this arrangement, the direction of hoisting could be reversed. The drum shaft was provided with a strap break, and there might be an arrangement by which a weight sliding in front of a signal board indicated the position of the kibble. Water whims were used mainly at the eastern mines—such as Fowey Consols and Wheal Friendship near Tavistock—where surface water was plentiful. The horse and water whim were introduced by the Harz miners.

The rotative steam engine was adapted to make a more powerful and faster hoist—the 'fire' or 'steam whim'. The first steam whim was Watt's engine erected at Wheal Maid in 1784, with a vertical spiral drum. A good example of a later type is preserved by the Cornish Engines Preservation Society, and can be seen beside the A.30 road between Redruth and Camborne. One early example of a steam whim of this type was that erected at East Wheal Crofty in the early nineteenth century, designed by J. Sims. This had a vertical beam engine coupled by bevel gearing to two 4 ft. diameter/horizontal winding drums,

TABLE SEPARATING FINE TIN FROM WASTE. The ore pulp is distributed onto the deck from the long box under 'Cousin Jack's' arm, flowing in through the pipe from the right. Water flows over the table from the perforated pipe running along the left hand, top, edge; the deck sloping down slightly from this edge to the bottom (right hand) edge.

The heavy cassiterite particles are thrown along the table, and remain close to the top edge, whereas the lighter waste tends to be washed down. The cassiterite can be seen separating as a light streak along the top of the table, discharging to the left. [1965]

FINAL DRESSING OF TIN ORE AT DOLCOATH in 1904, in round buddles of the convex type. The ore pulp flows off the cone in the middle into the round pit. The brushes slowly sweep over the settling particles to encourage an even deposit, the heavy ore material coming out first near the cone, the water and slimes overflowing through holes in the side of the pit. When the buddle is full, the ore-stuff is dug out in rings, according to grade.

A BELLOWS FOR ASSAY PURPOSES (dry tin assay) as used at Levant mine in the 1880's, and now preserved at Zennor Folk Museum.

VANNING SHOVEL. A tinstreamer using a vanning shovel to check the concentrate from a table. The cassiterite 'shows' as the light crescent near the tip of the shovel, which he has thrown up clear from the darker gangue material.

so arranged that the engine could hoist from two shafts at the same time (gearing was used to keep down the speed of hoisting and to aid in drawing heavy loads in the difficult shafts; many later whims had the bob direct coupled to a crank on the 'cage' or winding drum). Sheaves carried on frames directed and supported the chains to the shafts. The mouth of each shaft was protected by two hinged doors with a slit in the middle of each for the passage of the chains; the ascending kibble raised the doors, which fell back as soon as it had passed. The kibbles of wrought iron, holding from 5 to 7 cwt. each, were then tipped into waggons destined for the dressing floors. The later vertical cylinder steam whims, introduced about 1850, hoisted at up to 150 ft./min., using a steam pressure of 35 lbs. per square inch. Subsequently, powerful horizontal cylinder whims were installed, a particularly fine example being made by Messrs. Holmans in 1907 for the Robinson's Shaft at South Crofty—still in use until 1967 when replaced by an electric winder. This had two 22 in. by 4 ft. stroke horizontal cylinders, working at 125 lbs. per square inch, running at over 120 r.p.m. at full steam, and hoisting at over 2,000 ft./min. (including acceleration and breaking, lifting from the 335 fm. level to the landing brace at surface in less than one minute—a wind of 2,021 ft.).

These later horizontal steam whims were remarkably swift and smooth. The two winding drums were commonly about 8 ft. diameter. The steam exhausted direct to atmosphere with a deep throated "puff-puff, puff-puff". At the start of a wind with full steam on, the valvegear tinkled away merrily and, as the driver cut back on steam to decelerate the skips, the noise changed to a "buzz-buzz" as the Cornish double beat valves seated shut. The guiding pulleys and sheaves to the headgear rattled as the ropes shot over them, and the hoist could be stopped to the inch to 'land' the skip. The enginemen had great skill in controlling the hoist, and could move the rope less than an inch at a time. At the end of a hoist, with the engine at rest, the steam condensed in the valvegear with little "pop-pop" noises, and the enginemen waited for the tinging of the bell from the shaft landers to start the next skipload on its way. An example of a recent electric hoist is the one installed at Geevor in 1954, a B.T.H.-Wild winding engine, equipped with a 450 h.p. A.C. electric motor, with overspeed-overwind protection and Hobson brake retardation governor, hoisting at up to 1,500 ft./min.

Early shaft sinking was undertaken by the same methods as were used for winzing. The compartments for pump rods, rising main, and

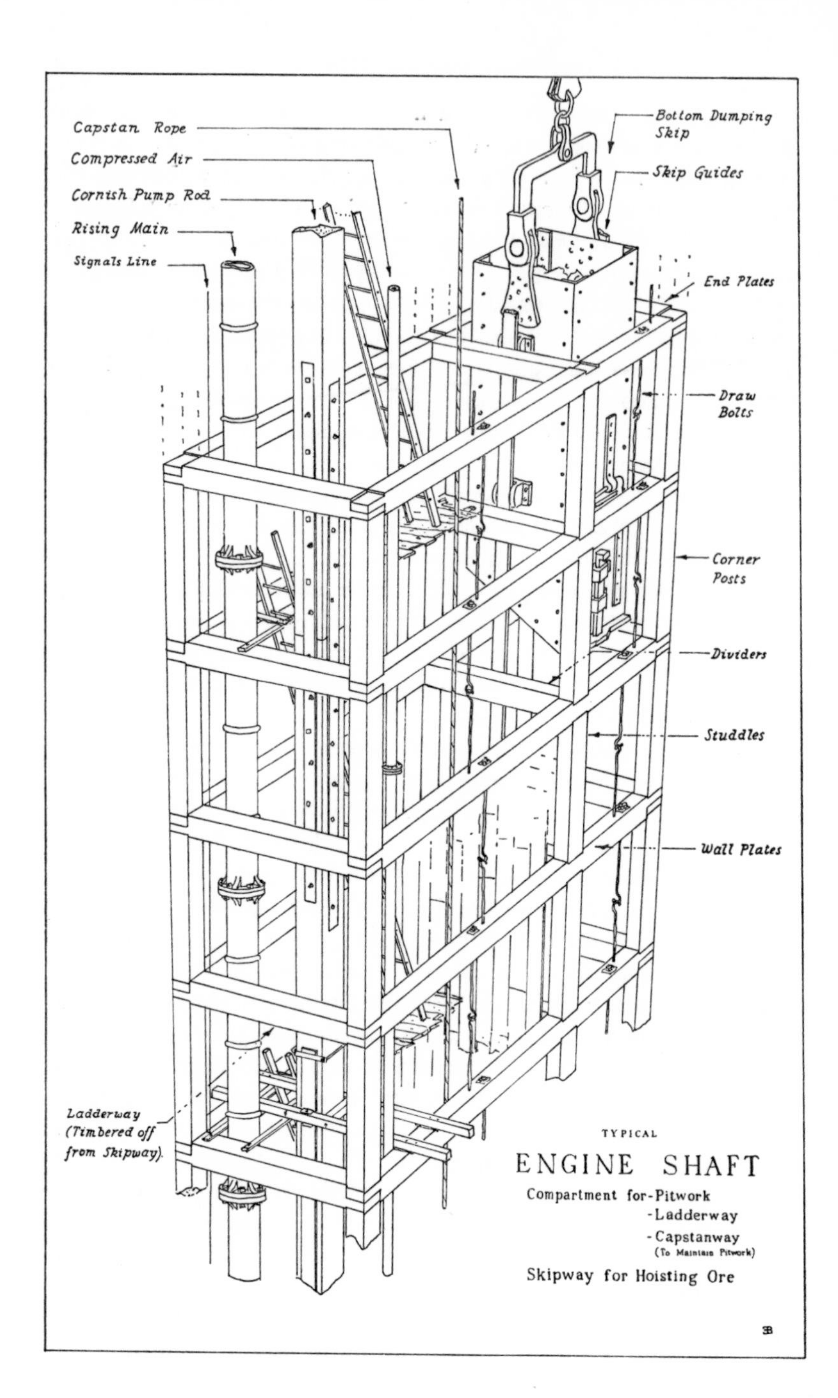
Capstan Rope
Compressed Air
Cornish Pump Rod
Rising Main
Signals Line
Bottom Dumping Skip
Skip Guides
End Plates
Draw Bolts
Corner Posts
Dividers
Studdles
Wall Plates
Ladderway (Timbered off from Skipway).
TYPICAL
ENGINE SHAFT
Compartment for-Pitwork
-Ladderway
-Capstanway
(To Maintain Pitwork)
Skipway for Hoisting Ore

access ladders, were timbered off from the kibble or skiproad by dividing boards, and later wooden guides for skips were added in the hoisting compartment. Movement of the hanging wall could damage shafts sunk on the lode and by the end of the nineteenth century shafts were located with more care, where possible being sunk truly vertical in footwall rock. The position was chosen so as to be able to develop the orebody and utilise the shaft most effectively over the estimated life of the working. The normal twentieth century Cornish shaft is rectangular, three compartment and finished to about 15 ft. by 6 ft. within the timberwork. Such shaftsinking requires a well planned work sequence. The drilling and blasting uses cuts as for driving a level. The rock is mucked out into kibbles, and then the timbers (pre-cut to size) are lowered and tightened into position by drawbolts and wedges against the rock sides. The timbers are cut with joints which will fit together tightly. A working shaft is commonly sunk deeper by using a small hoist in a compartment below the sump of the existing shaft. The new section is 'holed' and joined to the old when all is complete. A three compartment shaft of this type is adequate to cope with the average hoisting required to give an output of 80 tons black tin concentrate per month—large for a Cornish mine working the types of lode found in the West of England. Occasionally, larger four or five compartment shafts were sunk, such as the 1,461 ft. deep Allen's shaft, 19 ft. 6 in. by 6 ft. outside timbers, of Botallack mine which had five compartments and was completed in 1912. The extra compartments were for a capstan way for pump replacements, and an extra ladderway, but it is doubtful if such large shafts were necessary. William's Shaft at Dolcoath, started in 1895, 3,000 ft. deep, was unusual in that it was circular, of 17 ft. 6 in. diameter and brick lined, due to the poor ground encountered. The New Roskear shaft, also sunk by the reconstituted Dolcoath Company between 1923 and 1926, was 2,000 ft. deep, 16 ft. diameter and also brick-lined. This shaft, now part of South Crofty, was originally intended to locate a continuation of the lodes worked in North and South Roskear mines but the lodes cut were not as "kindly" as expected, in the country explored at the time.

Techniques in underground mining in the West of England have thus evolved over the years. As mines expanded, taking in larger 'setts' or ground boundaries, so improved mining practice evolved from experience. By the twentieth century, mining had become systematic, both in the development to keep 'ore reserves' prepared for extraction, and in stoping methods. Good vertical shafts were sunk, leaving shaft pillars of supporting ground where they cut through lodes. A combination of overhand and underhand stoping, as used today at Geevor gave high outputs and avoided problems of broken ore cementing together in overhand stopes due to slow drawing.

There has been considerable overlap in the evolution of mining methods, however. Some mines retained the old systems for many years after others had adopted the new, but a distinct pattern can be followed. It is instructive to compare the 3,000 ft. deep Dolcoath of 1895—the deepest Cornish mine—with present day metal mines in Africa, India and the Americas, some of which are now over 10,000 ft. deep. Many of the mining methods which are now used to work deposits abroad were evolved from those employed previously in Cornwall, particularly by the 'Cousin Jacks' who had to emigrate from their native county due to the mining slumps of the late nineteenth century. Underground mining still thrives in Cornwall at Geevor and South Crofty, and it is hoped that new mines will be opening shortly.

CHAPTER FIVE

ORE DRESSING

AFTER the ore had been won from the ground, further preparation was necessary to remove waste minerals before smelting. Such 'ore dressing' was a part of nearly all Cornish mining, to a greater or lesser extent, from the most distant past. The stream deposits—the earliest source—were rich and required only simple treatment, but with their exhaustion lower grade ores were exploited and these required more elaborate methods to separate the values from the waste. The higher the price that could be obtained for a metal, the more extensive this dressing could be, and the lower the metal content of the ore. For tin, the feed may now hold down to ¾% or less metal, but for the normally less valuable copper 2-3% had to be available to be worth concentrating, whereas iron ores were mined which would require only the minimum of treatment before being smelted.

The eventual metal content of the prepared concentrate varied from mine to mine, and frequently from month to month at a given mine. Thus a final 'black tin' (the concentrate for the smelter) commonly held about 60% metal. Apart from removing much of the gangue, particularly objectionable impurities which interfered with the quality of the metallic tin or copper, such as arsenic, had to be extracted although only present in small quantities. If the mine did not smelt its own ore and 'foul' concentrate was sold to a smelter, penalties would be charged against the price paid for having to work difficult material. In some mines, the lodes held several valuable minerals together, such as tin, copper and wolfram, which were separated and treated individually. The treatment area for the dressing was known as the 'dressing floors', or later the 'mill'; in the case of tin, the primary or 'green' concentration took place in the 'mill', and the final operations in the 'sweet' plant or 'tin yard'. At many mines, the ore was dressed in 'parcels'—heaps of ore tipped at the surface coming from different tributers, or allocated to individuals such as a mineral lord.

Until the nineteenth century the principles used in dressing remained virtually unchanged from the dawn of mining. The methods depended on the hand picking of values, the high density of the metallic ores compared with the waste (cassiterite sp. g. 6.8-7.1; chalcopyrite sp. g. 4.1-4.3; quartz sp. g. 2.65; chlorite sp. g. 2.65-2.94), and the ability to carry heavy grains of ore further against a stream of water than the lighter grains of the gangue of similar size. Recently, new principles to achieve separation have been used in addition, the ore stuff being extracted chemically, separated magnetically, or concentrated into a froth after selective sensitisation by physico-chemical surface effects.

Originally the dressing consisted of shaking up the sandy ore-stuff from the river beds as a 'pulp' with water in wooden bowls, so that the values could settle into a lower layer. The waste was then skimmed and washed off. Gold panning is an example of this technique. On a larger scale the ore was raked up against a stream of water in a channel cut in the ground or made of wood (a 'strake') with a bed of suitable material, such as turf, to catch the 'crop'. The best turf was considered to be that found about two miles to the east of St. Michael's Mount, in Marazion marsh, and was full of roots. The cassiterite tended to settle amongst the roots in a similar manner to the formation of a concentrate in a stream around stones and tufts of grass, the waste being carried on. When sufficient concentrate had been caught, the turfs were rolled up and shaken in a bowl about 2 ft. wide, provided with handles at the side, and the final clean black tin washed out.

By about 1500, a rough concentrate was made by leading the pulp into a shallow rectangular pit, with a boy stirring up the flow with his feet to aid the separation and distribute the material evenly. The heavy particles settled out first at the 'head', the deposit then being made up of progressively poorer values. The ore stuff was dug out, and straked according to grade. A preliminary concentration in this manner constituted an early form of 'buddling'.

The rich river sands and detritals of the first tinners could be dressed without crushing, but when pebbles, shoad stones and the lodes themselves were mined, the ores had to be crushed to liberate the crystals of the values locked up in the gangue. The early streamers and moor tinners powdered the coarse ore stuff by hand in their stone pestles and mortars, or on a larger scale by crazing mills. This latter consisted of a flat circular upper stone rotated by man, horse or water-wheel power about a vertical axis, worked on a lower fixed stone. The ore was fed in through a small hole in the centre of the upper stone, and the powdered material discharged around the rim. Crazing mills of this design were used in the Egyptian gold mines, probably before 200 B.C. The ground ore was then mixed with water and treated with the sands. The output from hand breaking and these mills was relatively small, and to treat larger quantities, stamp mills (probably introduced by the Harz miners) were adopted and put into use in Cornwall by the sixteenth century. The early stamps were made up of heavy iron-bound logs (stems) which were lifted and allowed to drop on to the dry ore-stones by means of tappets on the stem, and cams (or 'wipers') on a horizontal wooden barrel turned by a waterwheel. Carew in 1602 describes 3 and 6 head (number of logs) stamps in use in Cornwall which agree closely with illustrations in *Agricola* of Harz stamps. Later, it was found that the crushing was improved if the ore was kept wet. The fine pulp splashed out of the stamps, sometimes through a perforated plate sieve, and went on for treatment. The earliest stamps used a rectangular stone block as the mortar, about 4 ft. long by 1 ft. 6 in. square, which was turned to expose a new face as the heads wore into it. A fine example of such a stone can be seen at the Zennor Folk Museum. These early forms evolved into the Cornish Stamp, which had iron heads on wood and later wrought iron stems, working on to rammed stone: and later to the Californian stamp with improved cams and tappets, using a heavy steel head working on to a metal die.

By 1700 tin dressing methods had developed to a point where they remained unchanged, on some dressing floors, until the twentieth century. The ore was hand picked, broken by hammer into pieces no larger than a fist, and stamped. The stamp heads weighed about 140 lbs. apiece, and the grating put in front to size the pulp was a plate of metal approximately 1 ft. square, $^1/_{10}$ in. thick and punched with holes "the bigness of a moderate pin". The pulp discharged through the sieve and flowed through three pits in succession, where it settled according to density. The 'fore-pit' was the richest, the second and third held most of the remaining particles, whilst the 'slimes' and overflow ran to waste. The fore-pit stuff was treated in a 'square buddle', 7 ft. long, 3 ft. wide and 2 ft. deep. The dresser stood or sat on a board across the pit, spreading tin-stuff on a sloping surface at the head of the buddle (the 'jagging board') which had a series of small ridges parallel to the direction of a stream of water distributed over the slope from a water launder. The material was slowly washed into the pit, a broom being used to control the flow and encourage an even deposit, which was 'cut-out' in three or sometimes four sections when sufficiently thick—the material nearest to the jagging board—'the head'—holding the richest tin. The remaining portions were called 'first middle heads' (or 'crease'), second middle heads (or 'hind crease') and 'tails'.

The head was put into wood tubs (kieves) and stirred up in water with a shovel (tozing, tossing or terloobing) for about 15 minutes. The heavy tin particles settled down, and the low value 'skimpings' skimmed off the top. The rich bottom part was re-buddled to concentrate further, and again tossed. Finally the concentrate was 'packed' by smartly tapping the kieve with a hammer. 'Chimming' (introduced in the nineteenth century) was a modified form of tossing, where the kieve was inclined at 30° or so, and gave a clean separation with certain pulps, the concentrate packing into a smaller area. The first middle heads were treated in a similar manner.

The second middle heads underwent 'dilleughing' (dilleugh=to let go, or fly away). The dresser held a hair-bottomed sieve of coarse texture in a kieve two thirds full of water. An assistant then put in about 30 lbs. of ore and the sieve was then moved round and up and down, and from side to side, so that the small and light particles became suspended in the water, and by inclining the sieve were allowed to pass out and settle to the bottom of the kieve, the heavy tin remaining on the mesh. This action was aided by the sorting effect of the water coming up through the bottom of the sieve. The assistant then re-charged the sieve, and the process was repeated until sufficient concentrate had collected. The grade was checked with a vanning shovel, and if necessary re-dilleughed. The skimpings from the tossing kieves, and the material collected in the bottom of the dilleughing tubs, were re-buddled to crop out as much tin as possible. At some floors, the slimes discharging in the overflow as 'tails' or 'leavings' were treated on special 'frames'.

The tin-stuff settling in the second and third pits at the stamps was worked up in a different manner, being of finer grain size. The stuff was run into a circular pit (pednam) where it was stirred up, and gradually overflowed into a second pit about 10 ft. long, 3 ft. wide and 8 in. deep, divided into two sections by a wooden partition, that nearest the pednam being smaller and acting as a washing area. The head of the deposit which collected in the large part of the pit, was dug out for treatment on frames, and the poorer 'tail' part carried back into the small top section below the pednam, where it was again washed over ('trunked') for separation and dressing on frames—the tails from this second trunking being rejected.

The frames were made up of two flat wooden surfaces—the head and the body. The tin-stuff was spread over the head in ridges parallel to the direction of a stream of fresh launder water coming from a distributing arrangement similar to that used on the buddle. The water carried the ore gently forward over the slightly inclined body, where the heavy material settled on the surface and the light material flowed over to waste. The gentle flow, together with the large area of the frame, provided conditions where the fine tin could deposit and make a concentrate. When a suitable layer had formed, the frame was tilted up by hand and the crop washed into a wooden chest by a spray of water, and sifted. The sieving was done under water in a kieve, and took out the large grains of waste which had reported with the tin due to their weight. The sifted content of the kieve might then be re-treated on frames, tossed and kieved until a suitable grade of black tin was reached, and packed into sacks for smelting.

Poor tin ore was sometimes treated by 'tying' (or jigging). After a final buddling, the tin-stuff was put in a sieve—a typical example was 20 in. diameter by 5 in. deep, having a copper plate bottom perforated with holes 0.05 in. diameter at 0.4 in. pitch. This was given a series of short jerks in a tub of water, by hand, causing the heavy ore to settle through the pulp on to the perforated plate, and the waste skimmed off.

The details of the ore preparation used depended on the type of pulp that was being cropped—coarser 'sands' requiring a different technique from the fine 'slimes'. As the slimes were the most difficult to 'catch', much tin being carried over with the waste, stamping was limited by adjusting the coffer sieve size and the order of drop of the heads, so that the cassiterite crystals were freed from the gangue, but not excessively crushed. The exact details of the dressing varied from floor to floor, depending on the plant available and preference of the dresser. Checking the concentrates and tails at the various stages with a vanning shovel was an important feature of the operations. Such methods were still in use in the St. Austell area during the early 1900's.

Methods for dressing copper ores evolved with the increase of copper mining in the eighteenth century. Because the chief ore (chalcopyrite) was friable, breaking to a fine powder and so difficult to treat as it could be readily carried over in the tails, crushing was reduced as far as possible. Stamps were only employed on ores which were very hard, or disseminated. The ore was hand picked from waste, both underground and on the surface, so that milling costs would not be incurred on the treatment of low grades. After hoisting and tipping, the mine material was sorted into 'prills' (lumps of pure ore which did not require further concentration), 'drage' (ore mixed with gangue), and 'halvans', 'henn-aways' or 'leavings' (low values) which were stamped to release the ore, and treated by trunking, buddling and tozing, as for tin ore.

Muddy ore was washed on a 'griddle' (a coarse sieve) to expose waste pieces. The drage lumps were broken down, or 'ragged' by men using 10 lb. hammers, and the small pieces 'spalled' (broken finer) and 'cobbed' by girls. The cobbing broke off as much waste as possible from the ore fragments, and this was done with a special long-headed hammer. The stuff was then crushed, again by girls, or 'bal maidens', who used large flat-faced 'bucking' hammers working on square iron slabs. In 1796, John Taylor improvised a crushing machine ('Cornish

BAL MAIDENS 'COBBING' ORE

rolls') at Wheal Friendship in Devon from two discarded pump main pipes, which did not produce so high a proportion of fines as would have resulted if the ore had been broken by stamps. This early crusher was improved into an efficient machine consisting of two rolls of cast iron about 2 ft. diameter (one driven by a waterwheel), running against each other. The ore was fed between the rolls, the idler being pressed against the driven by a weighted crank lever, which allowed the free running roll to spring open if an extra large or hard lump was encountered. The crushed ore then passed through a revolving trommel (sieve), with a gauge of 6 holes/in^2, and the oversize that passed through was lifted up by a 15 ft. diameter 'raff wheel' and tipped back into the feed hopper for re-crushing. Rolls such as these at Devon Great Consols treated 40 to 60 tons a day.

The powder from the bucking sheds or rolls was then jigged. The material was put in a sieve, moved up and down in water, as for tying tin ore. The action of the particles of different density settling under 'hindered' conditions through the dense water-particle medium was particularly suitable for separating the relatively coarse copper ore pulp. After jigging for sufficient time, the light waste ('halvans') was skimmed off, along with any iron pyrite, by a 'limp'—a half circle of wood—the middle layer was taken for re-treatment by bucking and jigging, and the heavy copper values which had collected at the bottom

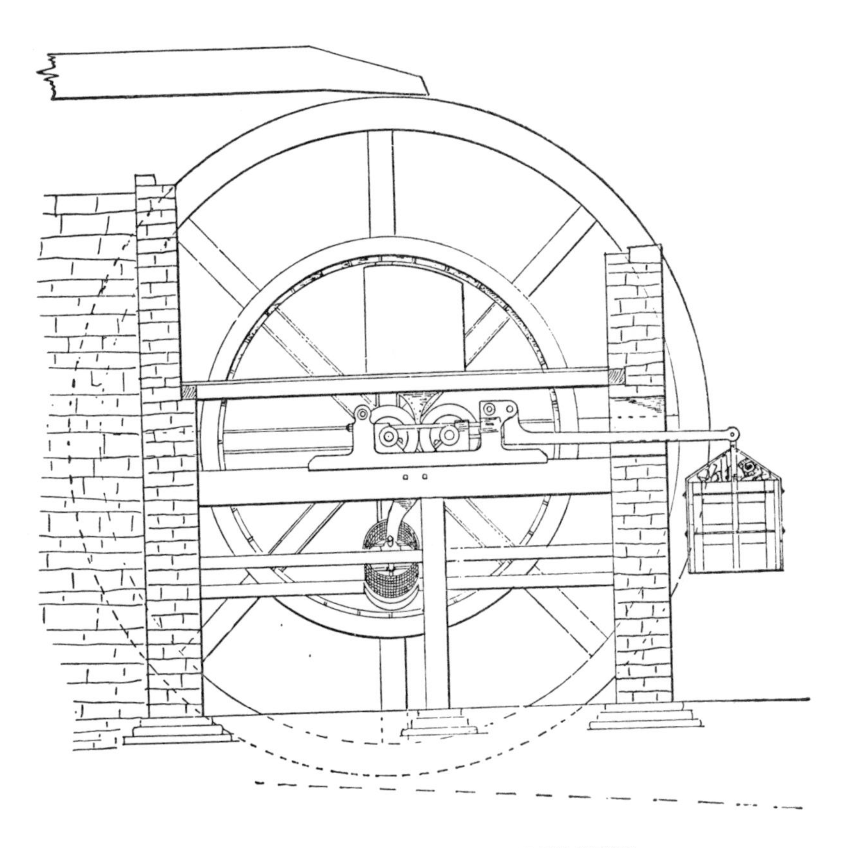

SECTION OF CRUSHER WORKED BY WATER-WHEEL

('ragging') tipped out. At first this jigging was done by boys bent double over the tubs, and holding the sieves, but later (by 1829) machines which could be worked by hand levers (often by women) were introduced. Waterwheel powered jigs were also constructed. Some fine ore-stuff passed through the jig sieve into the tub ('hutch work'), and was treated in buddles.

The copper period opened up new lodes, and large mines developed. As these lodes were worked down, many of the mines met tin ore, due to the zoning effect of the mineralization, and this resulted in a renewed expansion of tin mining later in the nineteenth century. Some of the mines did not convert to winning tin, as, under the cost book system of calls for money when capital was needed, there was a reluctance by short-sighted companies to make any large capital expenditure, notably the building of the more elaborate and expensive plant required for tin dressing. Moreover, many of the managers did not wish to adapt

JIGGING COPPER ORE

to the skills needed to work tin, having grown up with a 'copper outlook', and therefore obstructed any change. Copper dressing was predominantly by hand labour and tin dressing involved a greater use of machinery, including stamps, buddles and frames, with more careful control of the whole process. During the "copper years" many of the existing tin mines had improved their dressing equipment, and with the rise to importance of tin, those copper mines that installed tin floors did so on a large scale. The old tin preparation methods had been based on batch treatment, but the new floors were laid out more on flowline principles, although some of the detail operations were still done in batches. Compared with today's mills, they were confused and had the appearance of a shanty-town, but they constituted an important advance at the time.

The stamps, which were all-important in the initial preparation of tin ore, had evolved to the workmanlike 'Cornish' type and by 1770 were being built on an increasing scale. The old square buddle had been partly supplanted in some copper mines by the round buddle, introduced from Cardiganshire for the treatment of fines, and this was found to answer well for tin dressing. The efficient Cornish stamp, and the round buddle, formed the basis of the new tin floors.

An example of a Cornish stamp layout of the period was that erected in the early nineteenth century at Carn Brea mine, near Redruth. There were 72 stamp heads arranged in a line, of white cast iron socketed in wooden stems, secured by two iron bands. Each head was lifted five times per engine revolution, by cast iron, flanged, cylindrical cam barrels bolted together and driven by a steam engine in the middle through ratchets to allow the engine to run backwards when starting and establish vacuum, and couplings to allow for misalignment. The stamps were in sets of three and divided from each other by wood posts. Each head of a Cornish stamp usually weighed about 700 lbs. when new, and the fir stems were guided front and back by vees from the cross bars of the framing, and had a lift of 10 in. making fifty strokes per minute. The crushing was done in a 'coffer' of oak planks attached to the upright posts, on an 18 in. thick bed of quartz rammed in between masonry walls. The ore, hand ragged to 2 in. lumps, was fed in by waggons to 'passes' which led to the coffers. Water was run in with the ore, and carried the fine stamped particles through copper plates with 0.025 in. to 0.033 in. perforations, placed in front of the coffers. The maximum amount crushed per day was about 17 cwt. per head.

The pulp from the stamps was then often thickened by allowing it to settle, before passing to the buddles. A common arrangement for this was four 95 ft. long by 10 ft. wide pits, alternately being filled from the stamps, settled and drained, and feeding the buddles. The consistency of the de-watered pulp was adjusted by a controlled spray of water which washed the material to the buddles. Several types of convex and concave slope round buddles were designed. The type mainly used was of the convex form, made up of a wooden floor about 15 ft. diameter laid in a shallow pit. This had a cone at the centre which served both as a spreader and footstep for a vertical shaft with two arms carrying brushes which swept over the deposit. The vertical shaft was driven by an inclined shaft which had projecting knives working in a cylindrical mixer followed by a drum screen (8 holes per inch). The refuse from the screen feed was discharged to the side of the buddle, whilst the pulp passed to a funnel which spread the flow over the central cone and so down the inclined floor. The heaviest particles settled first and the lightest escaped at the circumference through holes in a sluice board provided with plugs at different heights. The brushes encouraged an even

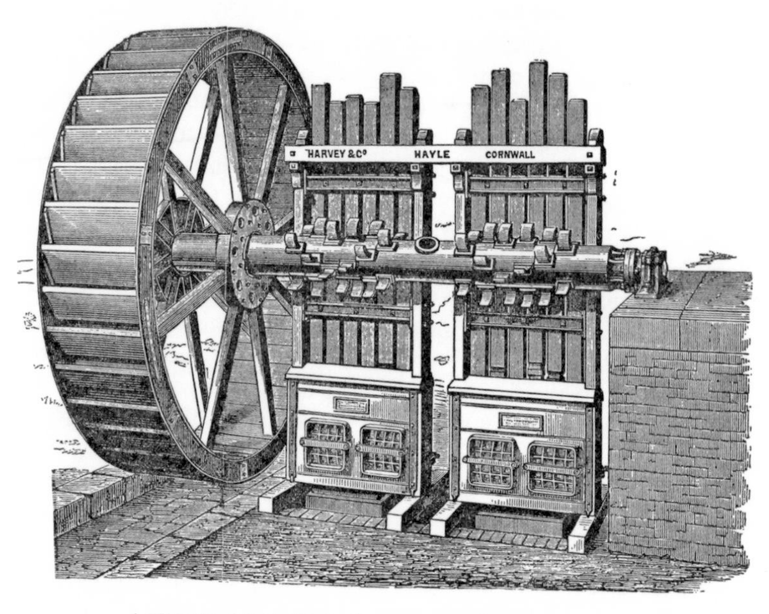

A TYPICAL SMALL SET OF (TWELVE-HEAD) CORNISH STAMPS
(Reproduced from an engraving in the catalogue of Harvey & Co., of Hayle, c.1880)

deposit free of guttering, and were counterbalanced by cord over pulleys to adjust them to the continually varying slope. When the deposit had reached a depth of 6-12 in., it was dug out in three or four concentric rings, that nearest to the cone being the richest.

Later buddles were of simplified design, made of cement-faced concrete, having a 4 to 6 ft. diameter central spreader, and had 3 to 8 brushes. The floor slope was 6 to 7 degrees and the brushes were hung loosely from the arms which made about 6 r.p.m. The mixing knives and round sieve were not used, the pulp being fed in by a launder through a perforated plate screen. These later designs could treat from 1.5 to 2 tons per hour; the finer the pulp being dressed, the larger the diameter of buddle used, 25 ft. or more being employed for slimes.

The concave buddles were made of wood and were commonly employed to concentrate the heads from the convex type. The pulp was distributed on to the sloping surface around the circumference by several revolving spouts fed from a central trough, and discharged through an opening at the middle with a raised ring having holes and plugs to control the thickness of the deposit and overflow. The tailings from the bundles contained fine tin in the form of slimes, and were treated either on the old hand tipped frames or the later automatic type.

The automatic 'rack' or 'rag' frames consisted of a fixed wooden table, holystoned flat, over which the slimes were distributed from a

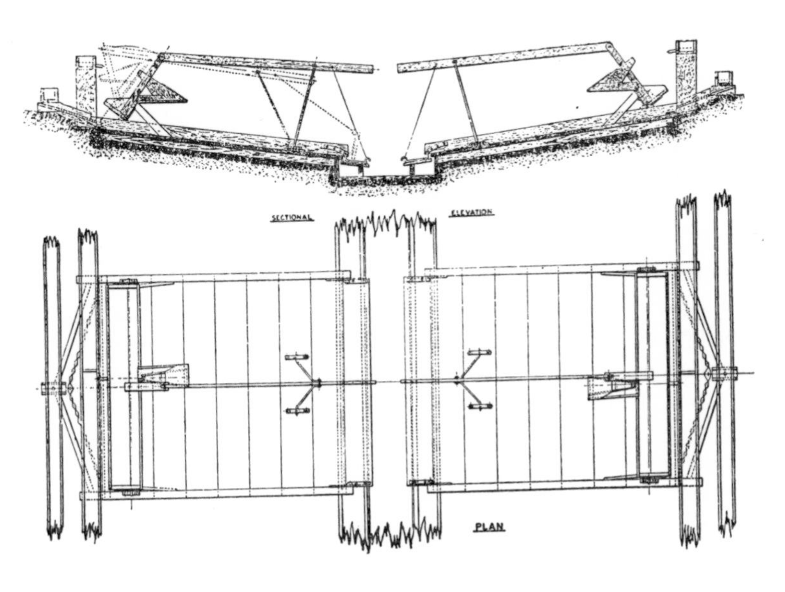

SELF-ACTING 'RACK' OR 'RAG' FRAMES
AS USED IN TIN DRESSING

trough, feeding a series of frames. As with the older design, the heavier particles collected on the table, and the lighter ones were carried over to waste. After a minute or so of flow, a fair amount of concentrate had settled on the table, and this was washed off by a cascade of water dashed on to the surface from a continually filled, self-righting trough, arranged to topple over when full of water. At the same time, an arrangement at the foot of the table connected to the toppling trough directed the concentrate into a special launder. These frames were also well liked by the tinstreamers cropping the waste discharged from the mines, who built them in large numbers. An improved version—used to this day—was the 'round frame'. The slimes were distributed on to a slowly rotating concave, circular wooden table, with a hole in the centre, over about two thirds of the circumference, and clear water gently flowed over from nearly all the remaining edge. The tails ran down into launders placed under the central opening, and the concentrate remaining on the deck was brushed and washed off continuously at the completion of a circuit, into the concentrate launder. The products from a frame could be re-framed individually to crop out as much as possible of the values, and waste which would not repay working-up finally discharged.

It was found that the ore could be buddled most effectively if the size distribution of the particles being treated was limited. Sieving large quantities of the fine particle size pulp would have been too slow, and 'classifiers' were devised which 'sorted' rather than 'sized' the material. An early classifier was the 'dumb pit'—a circular depression—where the pulp was poured in at the centre, the particles depositing in progressively smaller sizes from the feed point. The pit was then drained and the ore dug out to be buddled according to size. The later 'Cornish' classifier, first used by Capt. John Wilken in the Wendron mines in 1855, consisted of a conical wood box, with the apex pointing down. The pulp was fed into the top, and a jet of "hydraulic water" was introduced upwards from a compartment in the apex. The "hydraulic water" carried the light particles up, and caused them to discharge as an overflow, while the heavy particles settled down against the stream and below the jet, and reported out of a 'spigot' outlet at the bottom. Depending on the size of the classifier, the rate of the feed, and the amount of hydraulic water, the products could be controlled over a range of sizes, and, by using the classifiers in cascade, a split made into several ranges of size distribution which could be buddled separately as 'sand', 'fines', and so on. Due to the 'sorting' action, dense small grains tended to report with large light grains.

The development of the steam engine made power readily available, and encouraged the design of more elaborate ore dressing equipment during the latter half of the nineteenth century. This depended less on manpower, and could be run unattended for some time, as had been

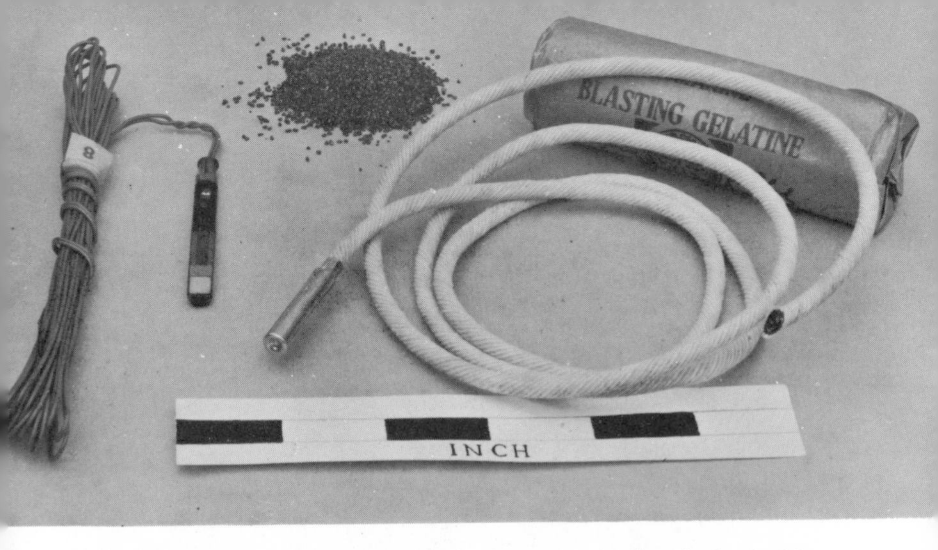

EXPLOSIVES. Sectioned delay action electric detonator, showing fusehead, delay element, initiating and base charge. Granules of gunpowder, showing the form most used in Cornish mines. The cartridge of blasting gelatine shows the parchmentized paper wrapper. The fuse, cut ready for lighting, is capper with a detonator, ready to make up the primer charge for a shot.

NING DIAL made by William ilton of St. Day about 1842. ote the scale with vernier reader r horizontal angles, and the ck and pinion on the vertical cle, also with a vernier reader, r measuring the underlie of clined workings.

TAYLOR'S SHAFT, EAST POOL AND AGAR. View showing the arrangement of the head-frame, pump house, boiler house, hoist and compressor buildings at Taylor's Shaft, built in the 1920's.
[1951]

POLDICE MINE. A general view looking towards St. Day village across the remains of the old arsenic zig-zag or 'lambreth' and its prominent stack. Cornwall is particularly rich in terms of industrial archaeology.

found possible with the tipping frames. 'Brunton's Belt', introduced in 1844 at Devon Great Consols for copper dressing, was an early example of a machine that could deal with large quantities of ore. A flat, endless canvas belt was moved up a slight, adjustable incline. About one third from the top a head-board delivered the pulp on to the surface, and somewhat above this a stream of water was distributed over the belt, The lighter waste was washed down the belt to a launder, and the heavier ore particles settled on to the canvas. The belt travelled slowly in a direction opposite to the stream, and carried the concentrate round into a tub of water below. During the latter part of the nineteenth century, the advantages of using such mechanical devices became apparent, both to meet the output of the mines and to reduce labour costs. The later Frue, Isbell and similar vanners used the same principle as the Brunton Belt, but were also given a side shake, whilst the belt was made of india-rubber. The vanner was found to be particularly suitable for treating the fine pulp which the stamps had to produce to free the cassiterite occurring in some of the mines, and which did not respond well to buddling. Large Frue vanner installations were put up in the Camborne district, such as that at Wheal Grenville built in 1901. The vanner could treat a rather wide range of mixed particle sizes as feed, but had the disadvantage of only giving a concentrate and tail, with no 'middlings'.

With the skill that had been acquired in ore dressing, it was realised that from some floors tin values were being carried away locked up in some of the coarser 'sand' sizes of waste. A new crushing device—the tube mill—designed by Michell and Tregoning, was put up at Wheal Peevor in 1880 to treat such ore/gangue 'middlings' particles in the circuit which could be separated from high grade and waste material during buddling. The mill consisted of a rotating cast iron barrel, partly filled with scrap iron, and the middlings pulp was fed in through one of the horizontal axles, crushed by the rolling scrap, and discharged through the axle at the opposite end. The degree of crushing in the early machines was controlled by regulating the feed. This was an important step in mineral dressing—today modified versions of the tube mill are widely used both to crush material in the mill circuit to release trapped ore, as at Wheal Peevor, and also as the main crusher, where they can be controlled by sieves in circuit to produce a smaller proportion of fines than stamps.

The shaking table, invented in 1844 by Rittinger and developed later by Wilfley, was adopted on some of the dressing floors in the late

1890's, having the advantage over the vanner that the discharge could be split into several grades. The table was made up of a slightly inclined rectangular, or similar plane surface of wood, about 15 ft. long by 5 ft. wide, sometimes covered with linoleum and small wood 'riffles', and given a jolting motion along the greater dimension by an eccentric drive. The riffles helped to guide the pulp which was fed on to the slightly inclined surface over part of the long, top edge and tended to flow down at right angles to the motion. Clear water from a perforated pipe or box flowed on to the table over the remainder at the top edge. The jerking movement threw the heavy particles of ore in the water film further along the long axis of the table than the light waste, and at the same time the washing water carried the gangue further down than the ore. The result was that the ore was separated and carried a greater distance along the deck in the direction of motion, so coming off the discharge end higher up than the waste particles. Middlings, or waste/ore mixtures, could be cut out by arranging suitable take-off troughs to catch the various products as they came over the edge. A table operated most efficiently with a feed of a rather more restricted size distribution range than that suitable for a vanner, and a 'classified' feed was better than a 'sized' feed from a sieve. The discharge from the spigot of a classifier cone was made up of small, heavy ore particles (which would throw furthest through the water film on the table deck), and larger grains of the lower density waste (which were washed down more readily by the water flowing across the deck). To make the most use of this property, classifiers (similar to those for buddles) were built in the floors which would deliver a range of products, from 'sand' to 'slimes', and the output from the individual spigots run to tables adjusted to treat the particular size distributions. The various concentrates were collected in wood boxes at the tables, and dug out to be carried to the next stage, or run to launders. In some cases the feed pulp was produced by tipping the thick deposit from a box into a wooden trough where a spray of water, adjusted to give the desired density of pulp, then washed the ore-stuff into a feed launder. Alternatively, the various grades coming off the tables were guided into the appropriate launder, which led to the next stage, thus using true continuous flow-line principles. This required careful adjustment of the water circulating in the mill, and might be impracticable if the run of mine feed varied widely in grade.

Tables are used extensively today for dressing tin ore in Cornwall, the slimes being passed to round frames. Round buddles and the tossing kieve are still important aids for achieving a final clean, high grade, black tin. Some of the sand ores, such as beach dredgings, are now given a preliminary concentration by running them down a Humphrey Spiral—a cast iron channel in the form of a vertical helix, the heavy product being cut out by suitable outlets cast in the trough, the centrifugal effect in the water stream achieving a separation. The concentrate is then tabled.

By the 1880's, the breaking of the run-of-mine ore was aided by using jaw crushers to prepare the feed for the stamps. In these, large stones were squeezed and broken between two hard metal plates, worked against the material by a suitable toggle linkage from an eccentric. Also at this time, the stamps themselves were being modified and the improved, 'Californian' type, developed on the Californian Goldfield, came into use, 40 heads being installed for instance at Dolcoath in 1892. These worked on the same principle as the Cornish stamps, but were more massive, and, with their improved lifting tappets and cams (which caused the heads to rotate and wear evenly—sometimes being called 'revolving stamps') used a head of 800 to 1,200 lbs., dropping against a metal die set in sand. They had a higher output than the Cornish type—on the hard ore at Dolcoath 1½ to 3 tons could be crushed per head per day. Pneumatic stamps, originally designed in 1871, such as the Husband and Holman, were installed at several mines by the turn of the century. These had a piston on the stamp stem that worked in a moving cylinder, suspended by means of two connecting rods from an overhead crankshaft, in which air was alternately compressed at the top and bottom ends. The air not only lifted the stem and attached head, but also drove it down on the crushing stroke with great force, the stamps working with a peculiar hissing sound that came in jerks.

Nowadays, for the initial 'primary' crushing, the large stones in the run of mine ore are rough broken in a jaw crusher placed either at the bottom of the shaft at the loading pocket (as at South Crofty) or in the mill. The discharge from this is then fed into a ball-mill (a tube mill loaded with steel balls) or a rod mill (loaded with steel rods) with a screen of about 20 mesh in series to control the size of particles in the discharge and avoid over or under milling for the first stage of concentration. The most effective dressings may be realised if the tin oxide is only partially liberated for the initial stages of treatment, finer crushing being delayed until the bulk of the values have been separated, and a middling cut made.

The spread of lode mining brought the problem of the removal of impurities such as arsenopyrite and pyrite, which interfered with smelting, and had not been present in stream tin. The earliest method

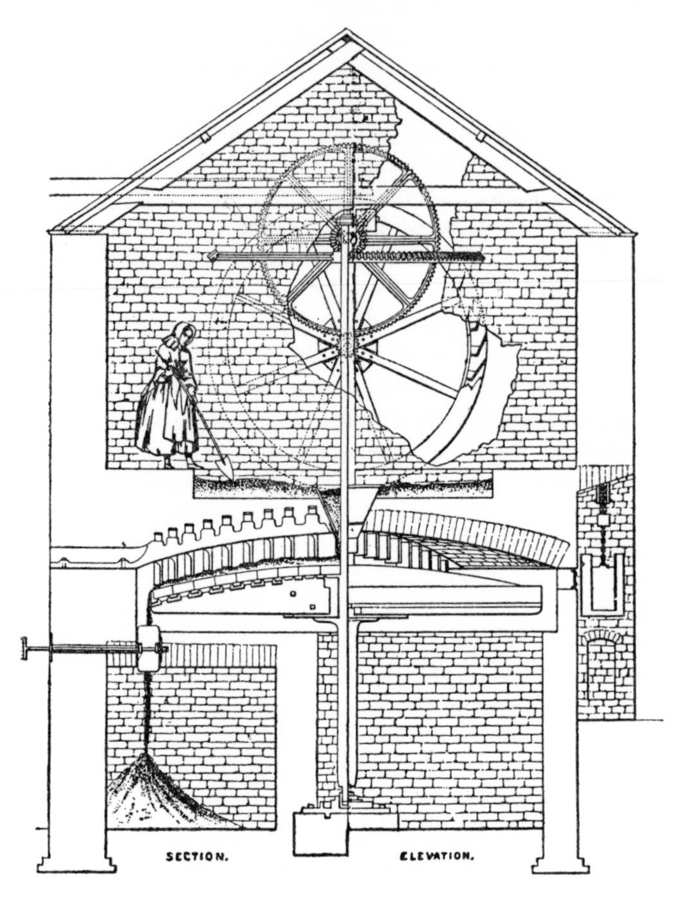

SECTION AND ELEVATION OF A BRUNTON FLAT-BED CALCINER

used for their elimination was to roast the ore in heaps, the FeAsS and FeS_2 being decomposed by atmospheric oxygen to arsenious oxide and sulphur dioxide, which were carried off in the fumes; the fine iron oxides remaining from the breakdown of these minerals were not as harmful, being frequently carried off in the slimes. Such burning or 'calcining' of ore appears to have become common practice in the eighteenth century. The Wherry mine, off Penzance, the ore of which was particularly 'foul', was notable for extensive burning of ore. Later, a partial concentrate ('whits') was made of the ore, holding approximately 30% tin with the impurities, and this was then burnt in batches of 7 cwt. at a dull red heat in a furnace 9 ft. long, of irregular width (3 ft. near fireplace, 5 ft. at widest point, and 1 ft. 6 in. at mouth, with an average height of 16 in.) with a chimney close to the mouth. The charge was introduced through a hole in the top of this hearth, and after heating up was hand raked until the fumes from the chimney "were no longer white" (i.e., until all the arsenic had been oxidised and driven off). The arsenic released in this manner was most damaging to surrounding crops and vegetation. Calciners became a common feature at all of the mines producing impure ores. By the 1850's, Brunton's calciner had replaced these simple furnaces, an early example being built at Wheal Vor in 1830. The whits were dropped continuously through a funnel on to the centre of a slowly revolving firebrick hearth of about 15 ft. diameter. Rabble arms were built into the roof over the moving hearth to scrape the roast slowly out to the edge, where it dropped down a discharge chute. Firegrates were placed in the calciner so that the charge was maintained at a red heat, and conditions adjusted so that sufficient air was drawn in and passed over the hearth to allow oxidisation to take place. A long flue connected the calciner to a labyrinth of chambers, where the arsenious oxide condensed as a 'soot', and was periodically dug out to be refined for sale. A typical flue was 1,000 ft. long, the arsenic crystallising between 140 ft. and 240 ft. from the calciner. Finally, the fumes passed out to a chimney. The roasted product then went on to be further buddled, tossed and kieved to form black tin. The last of these calciners worked until the 1950's, but the flotation process is now used to extract the impurities.

The flotation principle depends on the ability to make the surface of some minerals repel wetting by water, while permitting other minerals to be wetted. The minerals which repel wetting tend to concentrate from the pulp and attach themselves to an air/water interface, usually air bells blown in the pulp. They are then lifted out and held at the surface of the bubbles as a froth which is skimmed off. A few minerals tend to be naturally amenable to flotation, without requiring prior treatment, but in general it is found advisable to 'condition' the feed by adding chemicals which encourage a selective sensitisation of some of the constitutents of the pulp to be made by a suitable 'collector'. The 'collector' chemical—frequently a xanthate (or dithiocarbonate)—has a 'polar/non-polar' structure, one end of the molecule (polar) tending to attach itself to the surface of the particle of the mineral which it had the ability to coat (not affecting other minerals), the other (non-polar) end providing the non-wettable barrier. Only a minutely thin film is required on each particle to give it floatable properties. The process is adaptable, as under various conditions of acidity/alkalinity (pH value), and with different added chemicals to act as 'activators' or 'depressors' of particular minerals, the action of the collector may be modified so that a selective flotation may be made of individual constituents of an ore. Due to the very high surface tension of water, which will not permit the formation of the many small air bells that are most effective for achieving flotation, materials have to be added to allow suitable bells to be formed, and provide a froth which is sufficiently stable to hold the collected mineral, but not persist—breaking down when skimmed off. Pine oils and similar additives give good frothing properties to a pulp, but it may also be necessary to add other materials to stiffen or control the froth under certain conditions.

The principles of flotation were discovered in the 1860's, and during the twentieth century, it has become of major importance as the chief method used for the concentration of many ores, such as copper and zinc. Unfortunately, a suitable flotation process has not yet been developed for cassiterite, but the technique is of importance in the 'roughing' form to take out the arsenopyrite and pyrite, and also the copper frequently present in Cornish ores. Flotation is superior to calcining as it is more economical, particularly if the copper is recovered. The equivalent to the 'whits' for roasting are floated clean of these constituents in batch cells, taking a charge of 10-12 cwt. A charge is loaded into the cell, and conditioned by adding about 1 pint of oleum (sulphuric acid to control the pH), 200 c.c. creosote and 100 c.c. fuel oil (partial collectors and froth modifiers), 100 c.c. pine oil (frother), and 200 c.c. of a 15% solution of sodium ethyl xanthate ($C_2H_5O.CS_2.Na$—the collector). The pulp is then stirred up, and air forced in to create the bells (usually spun in through ports in the stirrer). After a while, the air bells carry up the arsenopyrite, pyrite and copper minerals, which form a black froth which is raked off into a tub. When no more material is floated, the contents of the cell, holding the tin values, is run out and passed on to the tinyard for final concentration. For cleaning coarse sand material, a modified 'skin' flotation is sometimes used on special tables (Holman-Michell table).

Tin dressing on the present day floors follows the earlier system of progressively separating the feed from the mine into a more and more concentrated 'black tin' on one hand, and reducing the values held in the waste from the various steps on the other. The equipment used is selected so that the most effective concentration can be made, considering the nature of the pulp, the particle sizes, mineral content and grade of values held. The degree of separation of values from waste that can be realised varies with the ore being treated, and is not complete. In 1900, it was estimated that up to 40% of the available tin in the ore mined was lost in waste, but at present the loss is well under half this amount. However, the quantities are sufficient to explain the continued existence of tinstreamers, who are still active treating the tailings from the two deep mines working today. Present Cornish milling practice varies somewhat depending on the feed—whether it is lode, detrital or re-worked material. In general outline (some of the stages may be omitted) the tin-stuff is fed to jaw crushers after washing and hand picking of waste rock, or crushed and waste eliminated by 'heavy media' separation (the light waste rock reporting to the surface of a heavy media pulp, such as ferrosilicon and water, the ore-bearing material settling down). The ore is then reduced further by ball mills, sometimes fed via an intermediate cone crusher, which discharge through vibrating stainless steel screens in the mill circuit (in place of the old perforated plates in front of the stamp coffers). After passing through mechanical rake classifiers to separate slimes, the pulp is separated into two size distribution ranges by hydrocyclones where the size split is made by spinning the pulp in a water vortex, and then classified to supply a range of spigot discharges for tables, the slimes passing to vanners and frames. The table middlings are re-ground by ball mills in circuit, and the partial concentrates obtained then have sulphide, arsenide and copper values extracted by flotation. The cleaned material is again tabled, vanned or put over frames according to size, and a final black tin, holding 60%-70% metal, worked up in the tinyard by using further tables or round buddles, tossing and kieving.

Entering a running mill, the first impression is of the innumerable streams of brown pulp shooting past in launders at all levels and directions. Rows of tables and vanners jog and sway quietly away with their sheets of pulp, and elevating dipper wheels (wheels with buckets around the rim used to rise the pulp for the appliances where required) are at work here and there. In the past, the stamps feeding the dressing floors roared and rattled and with the great Californian stamps the very ground shook underfoot, but ball mills do the same work now, turning with a steady rushing sound. Amidst these machines and the slowly turning frames and buddles one may find a mill man, whose skill shows in his apparent unconcern, all the stages being adjusted to run efficiently with the minimum of attention. When flotation is being used to clean the tin-stuff, the mill is pervaded with the peculiarly characteristic and pleasant smell of pine oil and xanthate.

Other techniques have been used for the dressing of Cornish ores, some only to a limited extent. Chemical extraction by solution along with solubility and crystallisation properties was applied to separate uranium and radium at South Terras mine. In several mines such as Giew, during the 1920's, native copper and copper minerals were leached out of tin concentrates by sulphuric acid, and later nitre-cake $Na_3H(SO_4)_2$. Flotation would be employed to do this today. Tungsten was extracted in the Calstock district in 1854 by heating the ore, mixed with soda, and lixiviating the roast. With the discovery of a rich body containing wolfram $(Fe.Mn)WO_4$, at East Pool in 1860, the tungsten was obtained as tungstate of soda by a similar process, the ore being treated with soda alkali in a reverberatory furnace. Tungsten was in increasing demand for steel alloys during the late nineteenth century, and was discovered in some tin lodes, particularly at Carn Brea, Tincroft and South Crofty. The high specific gravity of the wolfram (c.7.5) resulted in the ore reporting with the tin during the mechanical separation, and means had to be found to avoid this contamination. A new method was developed to separate wolfram from the tin concentrate, depending on the magnetic properties of the mineral. Examples were the plants installed at East Pool and at South Crofty, where the dried calcined ore was passed to a Wetherill Magnetic Separator. The feed was distributed on to a moving belt, which passed underneath electro-magnets. A cross belt, moving at right angles to the main one, ran under each magnet. The first and second poles were lightly energised, and caused iron minerals to be removed from the feed, and carried to the side by the cross belts. The third and fourth poles were more strongly energised, and attracted the wolfram, which was drawn out. Such a machine treated 10 tons per day of the wolfram bearing concentrate, the tin-stuff going on to the tin yard, being non-magnetic.

The majority of ores, other than tin and copper, were dressed by gravity methods. Lead and zinc were frequently treated by jigs and buddles, and due to the demand for cotton crop dusting and the consequent high price paid for arsenic, arsenopyrite was mined, particularly in some of the Devon mines, and dressed by jigging.

Ore dressing in the West of England has always required a considerable amount of skill, especially in the treatment of lode ores. The complex nature of such ores and the tendency for the cassiterite to be entrapped in the waste, brought problems as to how the tin could best be liberated, and also how objectionable minerals could be removed. In contrast, the Malayan cassiterite is relatively coarse and 'clean', and straightforward jigs are suitable for dressing. Many of the principles as used in the West of England dressing floors are now applied in other mining fields, such as gravity, chemical, magnetic, and flotation processes, although details differ and additional techniques may be required for the metals and ores treated—for example amalgamation—or a combination as necessary.

CHAPTER SIX

PROSPECTING, ASSAY AND SURVEY

ALONG with the main activities of winning and dressing ore, all the forms of mining required complementary skills, such as prospecting, assaying and surveying. Prospecting for ore carried with it a romance which acted as a great stimulus to mining—one has only to discover a small lode carrying values to experience this exhilaration. Once the characteristics of the ore and the deposits were appreciated, men could spread their search for riches and explore likely places, at first in the rivers and streams (which today may still show values) and later in the ground itself.

To help in their search, prospectors made use of many indications which could lead to a 'find'. 'Dirty' stained rocks, red cindery gozzan, changes in vegetation and material thrown up by burrowing animals could all hold clues. The stones of local walls and buildings, or those turned up during ploughing might contain some ore which, when interpreted, along with knowledge of the local geology, could enable a probable pattern of mineralization to be deduced.

The discovery of stones broken from the parent lode in the overburden (shoad) led to the early systematic search for the ore sources by 'shoading'. Pits were sunk to prospect the ground for the shoad stones, probably a development of the pitting by the old tinners, to determine the extent of their detrital values. Once the distribution was worked out the prospectors concentrated on trenching down to the rock bed in the area where the 'mother lode' was expected to be located, to find the back of the lode (costeaning). In the sixteenth century (early for lode mining) it was not considered unusual to follow shoad stones in this manner for up to five miles when searching for a deposit. Once the lode had been found, small shafts were sunk into it, and levels put out so that samples could be taken to test the extent and regularity of the ore-stuff. Cornishmen have long had a keen sense of the riches that might be held in their land, and were alive to any interesting finds. When excavations for foundations, roads or quarries were made, rock structures which had been covered by overburden were exposed, and new discoveries were sometimes made. New lodes were found during the driving of the extensive drainage adits for the mines, and in some one of the expected benefits was the possibility of finding new ore with the opening up of hitherto unexamined ground. Many of the great discoveries were made from surface indications, and the 'back' of out-cropping lodes could be readily prospected, but by the nineteenth century the majority of such deposits had been exploited. Lodes frequently do not outcrop to surface, and systematic cross-cutting underground became an important prospecting technique to examine the country rock on either side of the workings for parallel veins, particularly by mining companies who recognised the advantage of exploring for reserves while extracting a known lode.

The past history of mining in an area, especially of old burrows, could be of valuable help. By sampling earlier dumps, the nature of the ore and rocks could be studied and compared with any new discoveries. However, experience has shown that care must be taken not to be too optimistic that the strike and shoots of ore in an old mine will continue into unworked country—the 'old men' have usually mined most of the good ore. Plans and sections of abandoned mines were also frequently examined for information on the mineralization structure, but this had to be done with reservation as some were incomplete or inaccurate and, if prospecting underground, there was the risk of holing into old water-filled workings which were not recorded. When this was considered likely to occur, a pilot drill hole was maintained in advance of the face to tap into such water and give warning. Prospecting in depth today is now also accomplished by using diamond core drills, either drilling down from the surface (often across the strike at 45° underlie so as to intersect as wide a tract of country as possible and pick up indications), or from underground, being cheaper than driving cross-cuts and levels, which can be put out when favourable core indications are found. Core recovery is rarely complete, and it is only by actually examining a lode that reasonable judgement can be made of its possibilities.

In the past, many interesting phenomena, some fanciful, were credited with the power to reveal minerals. 'Lode-lights' or rippling white lights running across the ground at night (particularly noted at Stencoose, half a mile south of Wheal Music), were said to come from the back of a 'kindly' lode, and it is possible this was a form of phosphorescence. 'Dousing' was the term used to describe those with the ability to 'divine' ore with a suitable rod, or cone suspended on twine, made of a particular material to find a specific ore, which twisted or turned in their hands when held over a deposit. A friendly 'knocker' (gnome) could lead a miner to a rich bunch. Even today, the existence of these little people, some friendly and some not, is secretly half believed in by some of the tin miners.

Amateur interest in the geology and mineralogy of Cornwall, particularly during the eighteenth and nineteenth centuries, did much to improve mineral detection methods. Tests and systematic ways of examination were evolved, based on the crystal shapes, colour, behaviour when 'streaked' on an unglazed porcelain plate, reactions to heating in glass tubes and mixed with borax, and with a blowpipe on a charcoal block. Sections could be prepared in the laboratory for examination with a petrological microscope to determine the forms of the constituents of a specimen, and additional information gathered by seeing the colour changes peculiar to certain minerals when examined by polarised light. An invaluable 'tool' to prospecting is one of the textbooks which detail these tests, and cover a wide range of minerals, such as Collins' *Mineralogy of Cornwall and Devon*, (1871).

Added to these earlier tests, newer techniques are now applied. Geiger counters (sometimes placed in a survey aircraft) have helped in the search for radioactive ores in the West of England, as in other parts of the world. Magnetic and gravity anomalies have been studied to reveal rock formations. Portable filtered ultra-violet light sources are valuable when prospecting for minerals which fluoresce, such as some of those of tungsten. Very delicate spot reaction chemical tests can be used either to pick up the trace of a metal which occurs in the overburden covering a deep seated source, or to find high anomalies in an area which can indicate where more detailed prospecting should be undertaken. This technique has to be used with care in Cornwall due to the large number of spoil dumps which may contaminate the samples.

Determining the amount of valuable material in a sample, or assaying, was important in all the stages of mining—prospecting, winning the ore, dressing and marketing. The principal method used in Cornwall was 'vanning', other assays being relatively unimportant apart from the dry fire assay for concentrate. The vanning assay depended on the mechanical separation of the heavy valuable minerals from waste, and was done on a shovel, usually of special design. About one ounce of finely powdered ore-stuff was put on to the light, sheet-iron blade of the vanning shovel, and covered with water. The shovel was then swirled round, and any suspended mud then allowed to run off with the water.

More water was scooped on with the hand, and the swirling repeated until the sample was clean. Most of the water was then allowed to trickle off the edge, and after a few swirls to bring the material to the middle, the shovel was given a series of forward and upward flips ('vanned'), which threw the heavy particles in a crescent further up the blade than the lighter gangue. Careful manipulation washed down the waste into a central pool, where it was rubbed over with a hammer or flat section of steel shafting, to break up the particles and liberate any trapped values. It was then washed over, and again vanned, collecting the values into a single 'head' and ensuring all the recoverable ore had been thrown up. The remaining waste was then dropped off the shovel with a deft sideways flip. For the straightforward checking of ore content, a skilled vanner estimated both the quantity to put on to the shovel, and also the values by the appearance of the crescent he threw up. Underground, the miners used an ordinary shovel to van a sample to check the values and their positions in a lode.

In most cases the material to be vanned was made up of large stones, and had to be reduced and sampled. The ore-stuff was put on to a large plate of iron, about three feet square and ½ in. thick, let into the floor, and broken down to a coarse powder by a heavy flat-faced bucking hammer. This was then coned and quartered to reduce the bulk, and a portion powdered 'as fine as pepper' on an iron bucking table, with a smaller hammer which was used both to beat the ore, and also to grind it by being pressed down with one hand while worked around on the stones with the other. The quantity for careful assay was then either measured or weighed out from this.

Prior to 1860, the amount taken was a half-noggin wine measure full, scraped flat over the top, and the concentrate weighed in penny-weights—the system incorporating a factor which was considered to be approximately equal to the losses expected in the old time floors. After 1860 the charge was weighed out using a system peculiar to assaying. The '200' weight (the normal charge) was approximately one ounce, and the concentrate, or 'produce', was expressed as smaller a number (weight) such as '9½', which would represent about 4¾% of values.

After the head had been thrown up and the waste removed, the shovel was dried by holding it over a fire, and the concentrate brushed off into a clay crucible, in which it was roasted at red heat and stirred with an iron rod until sweet—or until the arsenic and sulphur had been driven off. This material was then re-vanned, dried, and the iron minerals extracted with a magnet before final weighing to find the produce. The vanning shovel is of Cornish origin, and although apparently simple by today's laboratory equipment standards, is a remarkably useful instrument, being adaptable to all gravity separable ores. The shovel may be used both for the detection and evaluation of ores, imitating the concentration techniques which would be used if they were mechanically dressed on a large scale, thus giving a close indication of the recoverable rather than total content of a sample. A skilled vanner could separate cassiterite, chalcopyrite and pyrite into three different 'heads' on a shovel, when examining a complex ore.

The highly enriched black tin for smelting could be more readily assayed by a 'fire' method. A weighed quantity of black tin (about 2 ozs.) was mixed with about 1 oz. of a reducing material, such as 'culm' (or powdered anthracite), and 4 dwts. of borax to aid the fusion and slagging. The mixture was put into a clay crucible and heated to a bright red in a bed of coals, or on the hearth of a small portable hand bellows, as used at Levant. After the charge had liquefied and settled down, the contents were tipped into an iron ingot mould, where the liquid tin metal which had been produced settled to the bottom and solidified as the 'lump', with a layer of slag on top. The slag was then separated from the metal on a sieve, and the lump weighed, the produce often being expressed in cwts. of metal per ton of black tin—thus 13½ produce referred to a concentrate holding that weight in cwts. per ton. The crude lump could then be re-melted in a small iron ladle, at a low temperature, the metal poured into a groove about 4 in. long by ¾ in. wide cut in a block of white marble, and the quality of the tin judged by the appearance of the cooled ingot (if clear and bright, the metal was of good purity), and the manner in which the ingot flattened under a hammer. When black tin was sold to a smelter, it was usual for three assays to be made—one by the mine, another by the smelter, and a third 'check' assay by an independent assayer to act as a referee in case of dispute. Originally, black tin was sold by the 'sack', each holding about twelve gallons beer measure (one gallon=about 282 cu. ins.), but is now sold by weight. Copper ore was fire-assayed with cream of tartar, the content expressed as produce, and was sold by a 21 cwt. ton.

The vanning assay is still used in Cornwall for routine sampling and control of ore dressing. A 'wet' assay (such as the Beringer) may now also be applied to determine the total tin content of a sample. The cassiterite in the roasted, weighed, charge is converted to metallic tin by heating strongly with zinc in a closed crucible, dissolved in hydrochloric acid, reduced by metallic nickel in a carbon dioxide atmosphere, and quickly titrated with iodine solution using a starch indicator. A 'Standard' solution of iodine is used, so that a simple calculation changes the burette reading to percentage tin. This assay is suitable for both low grade ore samples and concentrates, and gives the total tin held by the ore, which can be misleading as it may not be possible to recover all the tin present on the dressing floors. However, the assay is valuable to determine wastage and as an ultimate check. The amount of fluorescence, measured electronically, produced by putting a sample near a radio-isotope source, is now used to assay low grade samples.

Surveying—dialling—has always been of importance in mining for the fixing of mineral rights, boundaries, amounts and positions of ore mined, calculation of lords' dues and the control of the layout of the mine. The oldest known mining map, the Turin Papyrus of 1300 B.C., is of gold mines between the Nile and the Red Sea. The earliest surveying methods were simple and based on measured distances and instruments such as the dioptra. The angles and lengths of cords stretched down shafts and along the levels were noted, and by applying the geometric properties of triangles, the underground layout was reproduced full scale on the 'surveyor's field' at the surface. From this, ground boundaries were estimated, dues calculated, and the details found of the angles and distances required for development work. More elaborate instruments were needed as the mines grew deeper and the workings more complex. An early example of a compass adapted to mine surveying (an early form of 'dial') at Clausthal in the Harz is dated 1541, and consists of a 6½ in. diameter wood plate having several concentric circles filled with coloured wax, with a small compass in a brass box at the centre. The compass has a north-south line only, and bearings were recorded by "sighting" from this by placing the edge of the compass along a string stretched through the workings, and scratching on the wax. The distances between the survey points were measured with notched rods or ropes. The instrument was then taken to the surface and the survey laid out full size. Inclined sights were taken with a plummet line and graduated scale. Similar types of instrument came into use in Cornwall at about this time. Plotting on paper to a reduced scale from such surveys was a later development, and there are no known plans of underground workings until after the sixteenth century. Trigonometry, now widely employed to calculate positions from measured angles and distances, was not applied to underground work until the seventeenth century. Because of the system used for graduating some of these early compasses, Cornish miners often called north-south striking lodes "12 o'clock lodes". The unit measure of distance used almost universally in Cornish mines was the fathom (6 feet), but this has now been replaced by feet—although the level stations at South Crofty are still referred to in fathoms.

Such elementary techniques were adequate in Cornwall until the start of the great copper mining period, when the underground workings became large, extensive and complex. Surveying methods were developed so that positions could be set out to enable shafts to be sunk rapidly by mining from several different levels at the same time, the workings all meeting to form the shaft. The demand for greater accuracy resulted in the introduction of 60 feet (10 fathom) steel link chain measures, the first few links at each end sometimes being made of brass to avoid deflecting the compass needle, and stimulated instrument makers to provide accurate and robust surveying equipment suitable for mining—such as 'Lean's Dial', designed by Joel Lean, a Cornish Mine Manager. In particular, William Wilton, a "mathematical instrument maker", whose first workshop was a shed in his garden at St. Day, made many important advances, and also worked to a very high standard of precision, designing his own graduating machine to that end. Wilton improved the miner's dial (a compass made especially for use underground), and by 1841 in addition to the compass needle had made provision in his dials for surveying by 'fixed needle' (not using the compass)—carrying the survey forward, or 'traversing', by a system of measuring the angles between stations by backsighting and foresighting, starting from a known point, as used in surface surveying. The angles could be read to 2 minutes in azimuth by vernier, and also a removable level and quadrant was available for measuring vertical angles to 1 minute. Thus theodolite surveying principles could be used underground, and this made it possible to make more accurate mine plans and sections, and ensured that positions were known with greater certainty than from compass based surveys, with their attendant liability to have unknown deflections resulting from local magnetic influences. The Wheal Owles disaster of 1893, when 20 men were drowned in the Cargodna section, was the result of holing into the flooded workings of the abandoned Wheal Drea, because the compass surveys had not been corrected for the earth's changing magnetic declination over the years, and it was believed that the workings were well separated. Many of the old plans were the result of careful work, beautifully drawn and coloured, on paper-covered cloth or starched cotton. However, due to some of the plans being based on inaccurate compass surveys and measurements, great care has to be taken today when using them to obtain information, and, to add to these problems, the old records may be far from complete. Modern practice is to use a transit theodolite reading to 30 seconds of angle or better, avoiding compass surveys, and steel or metal reinforced tape measures. With this equipment, even under the poorest underground conditions, positions both in plan and section can be calculated to within inches over distances of many hundreds of feet.

Underground survey now includes the plotting of the nature and values of samples obtained by chipping out channels across the lodes at regular intervals. Sampling during the eighteenth and nineteenth centuries was rather haphazard, if done at all, and a variety of systems was used for reporting the results—sacks per ton, pounds sterling per fathom (which could only be interpreted for a lode of known width), 'produce' and latterly pounds per ton or a percentage. The plots of routine samples are used now to prepare a drawing sequence for the various stopes from day to day to ensure a steady grade of mill feed, and also make best use of the resources of the mine with the changing market price of metals.

CHAPTER SEVEN

CORNISH MINING ARCHAEOLOGY

THE WEST of England is exceptionally rich in material concerning the history of engineering and technology. The mining engineer can search and examine the clues remaining from the past for a fuller understanding of the background of mining, and those interested in technology can study the field of origin of a great deal of present-day engineering practice. The pattern of the stimuli, successes and failures which befell the people involved can be followed in the history of the mining booms and depressions.

The history of mining in the area stretches back to the unrecorded past. The varying outputs of metals reflect the relative scale of activity at different times, reaching their greatest peak in the mid-nineteenth century. The prices and demand for metals rose and fell, as at present, due to a number of factors, such as the Roman use of tin to make bronze, the need for bell metal for churches in the Restoration, for tin during the 'pewter period', copper during the Industrial Revolution, and tin once again (particularly for tin-plate) in the latter part of the nineteenth century. The opening and closing of other mining fields influenced Cornish mining—the Parys mines in Anglesey all but crushed the West of England copper mines, and later the availability of dredge-won Malayan tin affected the price of the metal so that deep mining became unattractive except for particularly well organised or rich mines. Today conditions are again favouring a renewal of Cornish mining.

The remains of the characteristic engine houses, dumps (or 'burrows') and old mine buildings are readily seen today in many parts of Cornwall, particularly in the western part of the county. Most of these date back to the mid-nineteenth century, the older remains having been either destroyed or replaced by the buildings constructed later, although only a small fraction of the total number remain, many having been dismantled for their building stones. Examples of the Cornish engine house may be seen, but any of the earlier Newcomen type are impossible to find. The typical massive stone structures are the relics of the homes of pumping and stamps engines and steam whims, with accompanying boiler houses. Added to these, the long flues and 'lambreths' (or labyrinths) connected to the square calciner houses at one end, and to a distant stack at the other, mark the site of the roasting and arsenic collection plants. The buildings are frequently laid out in similar groups—a large engine house for pumping at the side of the shaft, a smaller for a stamps engine, usually identifiable by the remains of the dressing floors in front of it, or the foundations for stamps at the side (such an engine might also have been employed to power the dressing floor machinery), and a further building to house a steam whim, usually in line with a shaft, and having masonry stands for a flywheel and cage. This was the typical assemblage of equipment which developed to operate the mines in the nineteenth century. As the mines expanded, further pumping engines and machinery were installed, often around new shafts spaced out along the lode, so that a row of buildings spread out at intervals may trace the strike of a rich lode of former days—such a line of engine houses can be seen at Killifreth, adjoining the Redruth-Truro road near Chacewater. It is interesting to note that the building of the boiler stack into the corner of the engine house dates back to the days of the earliest Newcomen engines, a practice which persisted with the majority of the houses up to the end of the steam period (Taylor's Shaft 90 in. pumping engine at East Pool and Agar, preserved by the Cornish Engines Preservation Society, has a separate stack). The remains of dressing floors can be seen in many localities, the relatively light foundations for the early Cornish stamps contrasting with the massive concrete required for the later Californian and pneumatic types. The round circles for the buddles may be quickly recognised, with the associated settling pits, and at the later mines the remains can be seen of the foundations for vanners and tables. The 'tailings'—dumps of fines, impounded by wall—are to be found at some sites, and the burrows of waste rock from either the hand picking or shaft sinking also serve as indications of where the 'old men' were busy. Such burrows along with the number of old shafts, are often the best indication of the scale of past activities. Due to the dressing methods used, copper mine burrows are usually more extensive than those from tin workings. Areas where such buildings and foundations can be found are along the cliffs near St. Just, on the southern side of Camborne and Redruth and along the valley near Bissoe. Other areas abound in material, but not to such a widespread extent. The remains of a few sets of stamps worked by a waterwheel usually dating from the 1880's contrast with the size of large mine layouts. Many of these little works may be found dotted along valleys (for example at Trelock near Nancledra and Trevaunance Coombe near St. Agnes) in the areas where tinstone could be picked from old dumps, or mined on a small scale at shallow depth.

The middle of the nineteenth century saw much excessive expenditure on surface installations by some companies, a practice which continued up to the start of the first world war. Money was lavished on unnecessarily elaborate equipment and ornamentation, which was sometimes installed before the lodes being prospected had been fully explored to determine their potential. An example was the last working of Botallack, where the great five compartment Allen Shaft was sunk between 1908 and 1912 in a doubtful section of the mine for future development potential. This was equipped with a large steam hoist (later moved to Taylor's shaft, East Pool) with a big compressor adjacent. An electric power station was built, with terrazzo flooring, and extensive dressing floors and workshops laid out—all in anticipation of great riches being cut into. This was typical of the period, when it was thought necessary to have these showpieces to demonstrate the resources of the company and impress those who were content to value investment prospects by such surface indications. This largesse was in contrast to the ornamentation bestowed on plant and machinery by proud designers and engineers, which amounted to a 'proper' finish to a good machine. In some cases, the woodwork making up the buildings and dressing floor appliances was attractively finished, and examples of this refined workmanship are now fast disappearing. Well proportioned and turned wooden balustrading was used in some of the offices, and important buildings of the mine, such as the whim and compressor house, frequently had spotlessly holystoned wood floors, embossed wallpaper, and carved mahogany instrument panels. Some of this 'mining architecture' is still preserved in the count houses of old mines, which have been converted to dwellings, or in the houses built for the mine officials in which the care lavished was in proportion to the status of the post, and might include an atttractively sited and landscaped garden with protective walls. These homes may still be seen in the principal mining districts, particularly at Camborne, Redruth and St. Day. The mineral lords—the owners of the mineral rights, receiving dues on any ore worked in their land—aided Cornish mining by using their influence to stimulate interest in their land, so that they could profit thereby. Lords fortunate in having rich lodes built fine estates, and vied with each other in their splendour. In some cases the mine buildings lay near their property, and had to be made suitably ornate to blend with the view—such as the stack of the old Carn Camborne mine which may be seen opposite the Pendarves estate near Camborne.

Other surface indications remaining include inclines for track ways, tramway cuttings and paths down the cliffs for the miners to reach a deep adit or for mules to take coal to a cliffside engine, such as the 'mule path' to the Crown's engines at Botallack. Numerous old shafts are left, many still open or only lightly sollared (timbered over), and these are usually surrounded only by a circular stone wall. Great care should be taken to avoid stumbling into such a shaft when looking over mine sites; in some cases, the mouth of the shaft is concealed only by a few brambles, all other indications having been swept away. The majority of the entrances to the adits, which are often found at the base of cliffs, have been blocked, but a few still remain open. These adits are now in a very poor state of repair in most cases, and may have sections where only rotting wooden stull timbers form the floor or roof, covering an old gunnis, so that exploration by those unaccustomed to mining can be most hazardous. Looking over old mine sites without someone with some knowledge of them can be dangerous, and one should always be prepared for the unexpected.

Many examples of tools and other equipment used in former days for mining are still in existence, and also the more specialised instruments, such as those used for assaying and surveying. These evolved with the developments in engineering practice, for example wooden shovels being replaced by wood tipped with iron, and then all iron. Some of the hand tools are peculiar to a district—this may be noted in the form the picks take as used in the St. Just area compared with those in the Camborne area. Collections of many varied tools can be seen at the Zennor Folk Museum and at the Holman Museum in Camborne. Although fast becoming scarce, the more specialised tools (such as those used for assaying) and other material—the resin/felt helmets and linen skull caps, candles, lamps, plans, sections and other documents, are still held by some families who were connected with mining.

In many instances the mines were stopped, and then became flooded, without recovery of the underground machinery and equipment. These now occasionally come to light when old workings are opened. Unfortunately, they are nearly always in poor condition as a result of the action of the corrosive mine water due to the breakdown of sulphide ores to form acid solutions, and in some cases chlorides from sea water which has seeped into the workings. The great network of adits to drain the mining areas are an important feature of these past workings, most adits having now fallen into disrepair, however. The scale of such ventures can be appreciated from the layout of the 'Great County Adit'

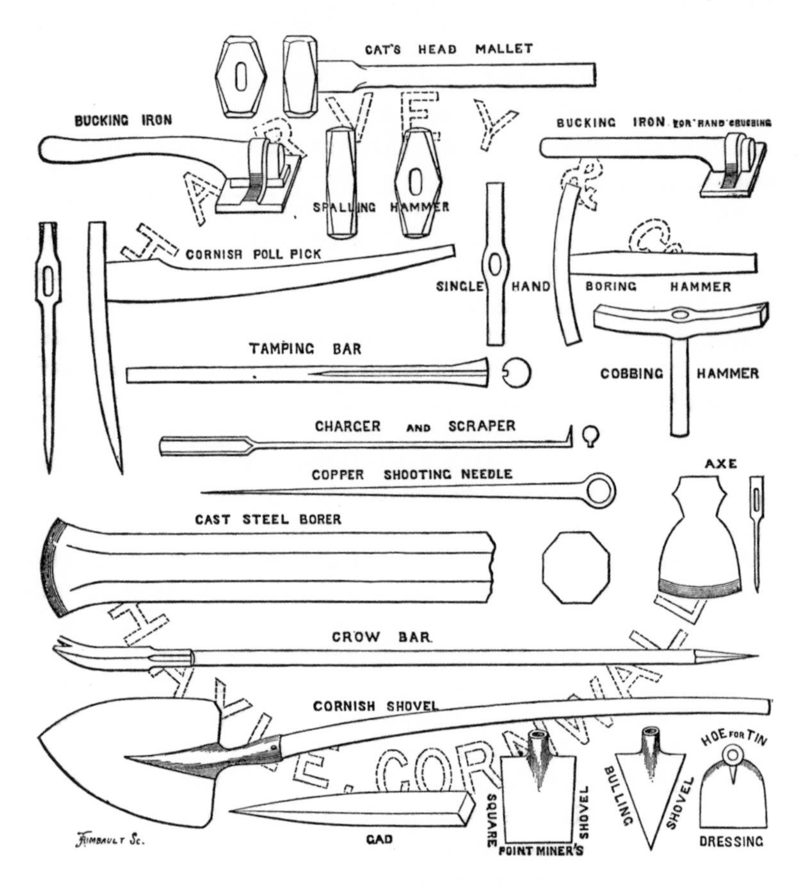

CORNISH MINERS' TOOLS
(reproduced from an engraving in the catalogue of Harvey & Co., of Hayle, c.1880)

which spread through the Gwennap district—an area of over 30 square miles was drained to a depth varying from 40-80 fathoms, by over 30 miles of tunnels which ramified under the area 'brought home' to the various mines. The extreme distance covered, from Cardrew mine near Redruth to the outlet near Bissoe Bridge, was well over five miles.

The ancillary industries which grew up with the mines also have an interesting history. In some cases, the old firms have adapted and changed to meet new needs, and are still active, such as Holman Bros., the famous engineering company, at Camborne, N. Holman & Sons of St. Just and Penzance, Harveys of Hayle, Bartles of Carn Brea, Visick of Devoran and Sara of Redruth. Others passed out of existence completely with the decline of mining, for example the once important Williams' Perran Foundry Co. at Perranarworthal. The remains of some of these old engineering works may still be found in several parts of Cornwall; for instance, the great machine and fitting-up shops to the left of the road leaving Hayle for Penzance, where Harveys made their engines. The instrument makers, Wilton of St. Day and later Camborne, Newton of Camborne, Jeffery of Camborne, Phillips of Tuckingmill, Dunkin & James of Penzance, and Letcher of Truro, form another group of practically forgotten craftsmen who contributed to mining, using their own ingenuity to design the specialised tools required to make these instruments. Candlemaking was also an important industry, and the remains of one factory at Par can still be seen.

The merchants who supplied the raw materials: candles, ropes, timber and coal, both profited from the mines and the miners as well as also commonly holding interests in various mining companies. Their scale of involvement and numerous dealings make a fascinating study when compared with present-day business practice. Transport during the heyday of mining was hampered by poor unsurfaced roads, and increased the remoteness of Cornwall both from the sources of raw materials, such as coal which had to come from Wales, and from the markets for the metals. Pack mules and horses were used for most carriage, while great horse teams, sometimes 40 or more, were harnessed up to haul engine bobs and cylinders from the foundries to their houses, or from mine to mine after a sale. Towards the close of the 'steam age' the engine bobs were so massive, some weighing forty tons or more each, that two steam traction engines coupled together were needed to pull such a load through the hilly parts of the Cornish countryside. Materials entering or leaving the county were usually brought by sea, and busy harbours grew up and flourished, of which Portreath, Devoran, Hayle, Trevaunance, Charlestown and Par are examples.

A feature of Cornish mining was the extraordinary degree of modification, adaptation and variety of uses to which the machines and equipment were subjected. Ore dressing plant was constantly altered, engines often changed hands many times, and were adapted from pumps to stamps and back to pumping engines, and many other duties. Parts of worn out machinery were modified to be used for some other purpose, for example a section of the scoggan catch of an engine being altered to make a stand for an auger drill. Tools were re-forged and shaped, and mine buildings changed without any inhibition. It is difficult to find any mining relic which is in the original condition in which it left its maker. Also, the workmen turned their hands to many trades, such as shaft-sinking, building headframes, installing engines, making offices, and the engineers on a mine would undertake an extraordinary range of repairs—the result of their efforts sometimes appearing crude, but making up for this in initiative and ingenuity.

Many lessons may be learnt by studying the factors—human, technical and economic—which influenced mining activities. Although the human and technical aspects have evolved over the ages, the localities and forms of the ore deposits remain the same, but a better understanding of their nature has come with experience. Also human frailties appear to have remained little changed—the terrazzo flooring of Botallack in the 1900's may be compared with some of the 'status symbols' of today. The invention of new machines and processes influenced mining methods considerably. The details of layout and techniques as used at present differ considerably from those of only half a century ago, and development work is now planned acknowledging this evolution, the life span of extraction in a new section being designed to last not more than thirty years, so that the introduction of future improvements will be less hindered by outdated planning.

The whole West of England mining region has a fascinating history going back many centuries, and much romance and interest may be found in it. It is instructive to try to visualise the scene during the great mining boom. Sir F. Bond Head in the 1827 *Quarterly Review* gives the following account, which accords closely with information which has been passed on by the old miners: "The situations (of the mines) . . . are marked out by spots . . . covered with what are termed 'the deads' of the mine—i.e., slaty poisonous rubbish, thrown up in rugged heaps, which at a distance, give the place the appearance of an encampment of soldiers' tents. This lifeless mass follows the course of the main lode (which, as has been said, generally runs east and west); and from it, in different directions, minor branches of the same barren rubbish diverge through the fertile country, like the streams of lava from a volcano. The miner being obliged to have a shaft for air at every hundred yards, and the stannary laws allowing him freely to pursue his game, his hidden path is commonly to be traced by a series of heaps of 'deads' which rise up among the green fields, and among the grazing cattle, like the workings of a mole. Steam-engines and whims (large capstans worked by two or four horses) are scattered about: and in the neighbourhood of the old, as well as of the new workings, are sprinkled, one by one, a number of small whitewashed miners' cottages, which, being neither on a road, nor near a road, wear, to the eye of the stranger, the appearance of having been dropt down *à-propos* to nothing . . . . Early in the morning, the scene becomes animated. From the scattered cottages, as far as the eye can reach, men, women, and children of all ages, begin to creep out; and it is curious to observe them all converging like bees towards the small hole at which they are to enter their mine. On their arrival, the women and children, whose duty it is to dress or clean the ore, repair to the rough sheds under which they work, while the men, having stripped and put on their *underground* cloths (which are coarse flannel dresses), one after another descend the several shafts of the mine, by perpendicular ladders, to their respective levels or galleries. As soon as they have all disappeared, a most remarkable stillness prevails—scarcely a human being is to be seen. The tall chimneys of the steam-engines emit no smoke; and nothing is in motion but the great 'bobs' or levers of these gigantic machines . . .

As soon as the men come *to grass*, they repair to the engine-house, where they generally leave their *underground cloths* to dry, wash themselves in the warm water of the engine-pool, and put on their cloths, which are always exceedingly decent. By this time, the *maidens* and little boys have also washed their faces, and the whole party migrate across the fields in groups, and in different directions, to their respective homes. Generally speaking, they now look so clean and fresh, and seem so happy, that one would scarcely fancy they had worked all day in darkness and confinement. The old men, however, tired with their work, and sick of the follies and vagaries of the outside and inside of this mining world, plod their way in sober silence—probably thinking of their supper. The younger men proceed talking and laughing, and where the grass is good, they will sometimes stop and wrestle. The big boys generally advance by playing at leap-frog; little urchins run on before to gain time to stand upon their heads; while the '*maidens*', sometimes pleased and sometimes offended with what happens, smile or scream as circumstances may require. As the different members of the group approach their respective cottages, their number of course diminish, and the individual who lives farthest from the mines, like the solitary survivor of a large family, performs the last few yards of his journey by himself."

The surface remains are but an indication where the chief labour and effort went on, beneath the surface. Most of the archaeology of Cornish mining lies buried in the closed, water-filled workings. Complex timbering, dams, 'bridges' over large gunnises, shaft pitwork, waggons, gigs, rails, chains, wood and iron pipes, and tools all still remain underground. It is there that the true scale of past labours manifests itself, testifying to the unparalleled skill of the Cornish miner.

INDEX

Books on Cornish mining and industrial history etc. . . .

CORNISH ENGINE HOUSES: A PICTORIAL SURVEY

H. G. ORDISH

A lavishly illustrated record of the finest of Cornwall's most distinctive industrial remains—the engine-houses that hallmark the landscape in the old mining districts. A second companion volume on the same subject is also to be published.

Small crown quarto . 70 *pages* . 77 *plates* . *paperback* . 15*s.*

HISTORIC CORNISH MINING SCENES UNDERGROUND

Edited by D. B. BARTON

Photographs taken underground in metal mines are rarities, particularly those taken before the turn of the century when photography under such conditions was difficult in the extreme. The historic value of the fifty plates in this book thus need little stressing, for they form a unique collection, complementing the earlier publication 'MONGST MINES AND MINERS reprinted in 1965. Lengthy captions explain for us these historic glimpses of the subterranean Cornish world that was peopled by tributers and tut-workers, sump-men and captains, and trammers and fillers.

Small crown quarto . 56 *pages* . 50 *plates* . *paperback* . 12*s.* 6*d.*

THE CORNISH MINER IN AMERICA

A. C. TODD

An important new book, of interest to both Cornish and American readers, telling the story of the great contribution to the mining history of the United States by emigrant Cornish miners—the men called Cousin Jacks. [United States edition published by The Arthur H. Clark Company, Glendale, California.]

Demy octavo . 289 *pages* . 24 *plates* . 7 *maps* . 50*s.*

ARTHUR WOOLF 1766-1837: THE CORNISH ENGINEER

T. R. HARRIS

A long-overdue biography of the man who, during the greatest era of the Cornish engine from 1810 to 1830, was the county's leading engineer, rivalling Richard Trevithick.

Demy octavo . 112 *pages* . 7 *plates and other illustrations* . 25*s.*

MINING FIELDS OF THE WEST

CHARLES THOMAS

A reprint of an early edition of this mining investors' and speculators' guide to the Cornish mines. Its author was a London mining sharebroker who had no connection with the Thomas family connected with Dolcoath mine.

Crown octavo . 96 *pages* . *paperback* . 15*s.*